Distributed Brillouin sensor in polymer optical fibers utilizing BOFDA

Distributed Brillouin sensor in polymer optical fibers utilizing BOFDA

Der Fakultät für Elektrotechnik, Informationstechnik, Physik
der Technischen Universität Carolo-Wilhelmina zu Braunschweig

zur Erlangung des Grades eines Doktors

der Ingenieurwissenschaften (Dr.-Ing.)

genehmigte Dissertation

von Andy Schreier
aus Bad Saarow

1. Referent: Prof. Dr. rer. nat. Thomas Schneider
2. Referent: Prof. Dr. Alayn Loayssa

Eingereicht am: 31. Oktober 2019

Mündliche Prüfung am: 04. Dezember 2019

Berlin, 31. Oktober 2019

Bibliografische Information der Deutschen Nationalbibliothek
Die Deutsche Nationalbibliothek verzeichnet diese Publikation in der Deutschen
Nationalbibliografie; detaillierte bibliografische Daten sind im Internet über
http://dnb.d-nb.de abrufbar.
1. Aufl. - Göttingen: Cuvillier, 2020
Zugl.: (TU) Braunschweig, Univ., Diss., 2019

© CUVILLIER VERLAG, Göttingen 2020
Nonnenstieg 8, 37075 Göttingen
Telefon: 0551-54724-0
Telefax: 0551-54724-21
www.cuvillier.de

1. Auflage, 2020
Gedruckt auf umweltfreundlichem, säurefreiem Papier aus nachhaltiger Forstwirtschaft

ISBN 978-3-7369-7146-2
eISBN 978-3-7369-6146-3

Abstract

In this thesis, a distributed Brillouin sensor in perfluorinated polymer optical fibers utilizing BOFDA is presented. These commercially available polymer fibers offer beneficial characteristics for sensing applications such as higher break down strain up to 100 %, minimal bending radii below 2 mm, higher sensitivity to temperature and lower sensitivity to strain compared to their silica equivalent. However, polymer fibers are multimode fibers. In multimode fibers the process of Brillouin scattering becomes increasingly complex as generally many optical and acoustical modes are involved. The underlaying process of stimulated Brillouin scattering (SBS) in graded-index multimode fibers is presented and its theoretical limitations are outlined.

The dominant excitation of the fundamental mode is key to generate SBS in perfluorinated polymer optical fibers. Combined with the modal selection for data evaluation, a fiber sensor based on SBS can be designed. For this purpose boundary conditions for connection misalignments are estimated utilizing theoretical coupling efficiencies. Hand-crafted fiber interconnections demonstrate significantly lower insertion losses compared to previous publications.

The SBS parameters - backscattering power, linewidth and frequency shift - are inclusively related to the environmental parameters humidity, temperature and tensile strain. Furthermore, the perfluorinated polymer optical fibers are characterized with respect to the influence of relative humidity and temperature changes on spectral transmission absorption and Rayleigh backscattering, respectively. The obtained results demonstrate that the humidity-induced Brillouin frequency shift is predominantly caused by the swelling of the fiber over-cladding that leads to fiber straining. The determined analytical description of the Brillouin frequency shift serves as a basis for the distributed fiber sensor.

The chosen wavelength of operation at 1319 nm corresponds to lower fiber propagation loss ($<$ 37 dB/km) compared to other approaches at 1550 nm (150 - 250 dB/km). A 86 m PFGI-POF was successfully measured by BOFDA with spatial resolution of 3.4 m. Thus, the sensing range is demonstrated to be extended by a factor of four and the number of sensing points is increased by a factor of five. A non-strained fiber segment is successfully detected within a strained sensing fiber. However, the full potential of the proposed setup with equipped optical filters remains unclear.

The findings related to humidity influences can serve as a basis for future distributed humidity sensors not only limited to stimulated Brillouin backscattering.

Zusammenfassung

In dieser Arbeit wird ein verteilter Brillouin-Sensor in perfluorierten optischen Polymerfasern unter Verwendung von BOFDA vorgestellt. Diese im Handel erhältlichen Polymerfasern bieten vorteilhafte Eigenschaften für Sensoranwendungen wie eine höhere Durchbruchdehnung von bis zu 100 %, minimale Biegeradien unter 2 mm und eine höhere Temperaturempfindlichkeit bei einer geringeren Dehnungsempfindlichkeit im Vergleich zu Quarzglasfasern. Polymerfasern sind jedoch Multimodefasern. In Multimode-Fasern wird der Prozess der Brillouin-Streuung zunehmend komplexer, da die Anzahl der beteiligten optischen und akustischen Moden stark ansteigt. Der zugrunde liegende Prozess der stimulierten Brillouin-Streuung (SBS) in Gradientenindex-Multimodefasern wird vorgestellt und seine theoretischen Grenzen aufgezeigt.

Die dominante Anregung der Grundmode ist der Schlüssel zur Erzeugung von SBS in perfluorierten optischen Polymerfasern. In Kombination mit der Modalauswahl zur Datenauswertung kann Fasersensor auf Basis von SBS entworfen werden. Zu diesem Zweck werden die Randbedingungen für maximale Versatzdistanzen unter Verwendung theoretischer Kopplungseffizienzen abgeschätzt. Handgefertigte Faserkupplungen weisen im Vergleich zu früheren Veröffentlichungen signifikant geringere Einfügungsverluste auf.

Die SBS-Parameter - Rückstreuleistung, Halbwertsbreite und Frequenzverschiebung - sind eine Funktion der Umgebungsparameter Feuchtigkeit, Temperatur und Zugspannung. Darüber hinaus sind die perfluorierten optischen Polymerfasern hinsichtlich des Einflusses von relativen Feuchtigkeits- und Temperaturänderungen auf die spektrale Transmissionsabsorption und Rayleigh-Rückstreuung charakterisiert. Die gewonnenen Ergebnisse zeigen, dass die durch Feuchtigkeit induzierte Brillouin-Frequenzverschiebung hauptsächlich durch das Quellen des Faserüberzugs verursacht wird, was zu einer Faserdehnung führt. Die ermittelte analytische Beschreibung der Brillouin-Frequenzverschiebung dient als Grundlage für den ortsverteilten Fasersensor.

Die gewählte Betriebswellenlänge von 1319 nm entspricht einer geringeren Faserdämpfung im Vergleich zu anderen Ansätzen bei 1550 nm. Ein 86 m lange PFGI-POF wurde erfolgreich mittels BOFDA vermessen mit einer Ortsauflösung von 3,4 m. Dieses messtechnische Resultat erhöht die Messreichweite um den Faktor vier und die Anzahl der Messpunkte um den Faktor fünf im Vergleich zu früheren Aufbauten. Das volle Potenzial des vorgeschlagenen Aufbaus bleibt jedoch Gegendstand zukünftiger Untersuchungen.

Die gewonnenen Erkenntnisse der Feuchtigkeitseinflüsse können als Grundlage für künftige ortsverteilte Feuchtigkeitssensoren dienen, die sich nicht nur auf die stimulierte Brillouin-Rückstreuung beschränken.

Acknowledgments

First, I would like to express my sincere gratitude to Prof. Dr. rer. nat. Thomas Schneider. He patiently taught me the theory of the SBS process in single mode silica fibers. Without this essential knowledge, I would have struggled developing a deeper understanding of the SBS process in multimode fibers with several acoustical modes. Furthermore, he taught me how to think about problems to develop multiple solution approaches. Without his support this thesis would not have been possible.

I am grateful for Prof. Dr. Alayn Loayssa examining this work. It is pleasure for me having an examiner with more than a decade experiences in cutting edge distributed Brillouin fiber sensing.

Moreover, I would like to thank my former colleagues at BAM for their support, the help in the laboratory and the stimulating discussions in the field of fiber optic sensing. A special thanks to Dr. Katerina Krebber, who paved the way for my PhD project. In particular I would like to thank Dr. Aleksander Wosniok for his encouragement and inspiration with the last years. I am grateful for his patience in guiding me through the theory, the laboratory implementation, the data evaluation schemes and the error propagation of the BOFDA technique. Dr. Sascha Liehr and Prof. Dr. Stefan Kowarik for fruitful discussions and improving my skills in writing scientific papers. I would also like to thank Thomas Kapa and Marcus Schukar for their active support in the laboratory. It was an honor for me to work with the entire team.

Finally, I would like to express my true and deep gratefulness to my family, friends and girlfriend Steph for their moral support and understanding over the last few years.

Contents

A	amplitude of the acoustic wave
A_{eff}	effective illuminated area
a	core radius of a fiber
B	bandwidth
b	magnification factor
B	experimentally determined Sellmeier coefficient
BFS	Brillouin freqency shift value in empirical description
BFS_0	theoretically Brillouin freqency shift value at 0 °C, 0 %r.h. and 0 % strain
c	velocity of light in vacuum
C	experimentally determined Sellmeier coefficient
$c_{\hat{p}p}$, $c_{\hat{q}q}$,	coupling coefficient of field amplitude distributions
$C_{h,abs}$	absolute humidity Brillouin frequency shift coefficient
$C_{h,rel}$	relative humidity Brillouin frequency shift coefficient
$C_{\hat{p}\hat{q}pq}$	complex cross field amplitude
C_S	strain Brillouin frequency shift coefficient
C_T	temperature Brillouin frequency shift coefficient
CHE	coefficient of hygroscopic expansion
CTE	coefficient of thermal expansion
E	electrical field
F	electrostrictive force
f	frequency
f_B	Brillouin freqency shift value in analytical description
f_D	frequency difference between pump and probe wave
f_{err}	frequency uncertainty within a confidence interval of 68 %
f_m	modulation frequency
f_{pol}	sampling frequency of the polarization scrambler
f_{step}	frequency increment / step width
f_{uncert}	frequency uncertainty within a confidence interval of 99.5 %
g_B	Brillouin gain factor
g_p	peak value of the Brillouin gain coefficient
$g(t)$	impulse response as a function of time
$g(z)$	impulse response as a function of distance
$G(jf)$	transfer function
h	humidity

H	Hermite polynomial
I	optical intensity
i	electrical current
K_B	polarization factor, $K_B = 1/\eta_P$
k	wave number, $k = 2\pi/\lambda$
l	length
L	length of fiber under test
L_{eff}	effective length of the fiber under test
L_{max}	maximum length of fiber under test
m	auxiliary variable for c_{00}
m_P	degree of modulation for the pump wave
M	number frequency steps to obtain $G(j\omega)$
n	refractive index, $\Re\{n\}$:group index
P	polarization
p	counting variable
p_{12}	longitudinal elasto-optic coefficient
q	counting variable
r	spatial variable perpendicular to ϕ and to the optical axis of the fiber
R	radius of curvature of a Gaussian beam
R_p	power reflection coefficient
R_{ph}	responsibility of a photodetector
RBW	resolution bandwidth
S	strain
t	time
t_d	pulse duration
T	temperature
V	fiber parameter
v_a	velocity of sound
x	spatial variable perpendicular to y and to the optical axis of the fiber
y	spatial variable perpendicular to x and to the optical axis of the fiber
z	spatial variable along the optical axis of the fiber
α	optical loss coefficient
β	spatial propagation constant of a wave, $\beta = 2\pi z/\lambda$
ΔBFS	uncertainty of the Brillouin frequency shift
Δf	frequency difference
Δf_B	Brillouin linewidth / full width at half maximum of Lorentzian
Δh	change of humidity
Δn	refractive index difference
ΔT	change of temperature
Δz	spatial resolution
$\Delta \varepsilon$	change of permittivity
ε	permittivity
ϵ	elastic modulus
ζ	linear factor
η	noise function
η_P	polarization mixing efficiency
θ	angle between two wave fronts or fiber facets
κ	power coupling coefficient

λ	wavelength
λ_a	wavelength of the acoustic wave
λ_P	wavelength of the optical pump wave
Ξ	pulse broadening
ρ	density
σ	standard deviation
χ	susceptibility
ψ	transversal field distribution
Ψ	phase of wave / oscillation
ϕ	spatial variable perpendicular to r and to the optical axis of the fiber
Φ	Gouy Phase of Gaussian beam
ω	Gaussian beam diameter in the x-y plane
ω_0	mode radius of a fiber
∇	Nabla operator

ASE	amplified spontaneous emission
BFS	Brillouin frequency shift
BGS	Brillouin gain spectrum
BOCDA	Brillouin optical correlation domain analysis
BOCDR	Brillouin optical correlation domain reflectometry
BOFDA	Brillouin optical frequency domain analysis
BOFDR	Brillouin optical frequency domain reflectometry
BOTDA	Brillouin optical time domain analysis
BOTDR	Brillouin optical time domain reflectometry
CHE	coefficient of hygroscopic expansion
CTE	coefficient of thermal expansion
CYTOP	cyclic transparent optical polymer, poly(perfluorobutenylvinylether)
DCT	discrete cosine transformation
DMD	differential mode delay
EDFA	Erbium-doped fiber amplifier
ER	extinction ratio
ESA	electrical spectrum analyzer
FBG	fiber Bragg grating
FC/APC	fiber connector with angled polish contact
FoM	figure of merit
FPI	Fabry-Perot interferometer
FSR	free spectral range
FUT	fiber under test
FWHM	full width at half maximum
GI	graded-index
GI-MMF	graded-index multimode fiber
GRIN	graded-index
iFFT	inverse fast Fourier transformation
iOFDR	incoherent optical frequency-domain reflectometry
L-BOFDA	linear configured Brillouin optical frequency domain analysis
LMA	Levenberg-Marquardt algorithm
MMF	multimode fiber
MZM	Mach-Zehnder modulator

Nd:YAG	neodymium-doped yttrium aluminum garnet
NMF	noise model function
OCDR	optical correlation-domain reflectometry
OFDR	optical frequency-domain reflectometry
OSA	optical spectrum analyzer
OTDR	optical time domain reflectometry
PC	polarization controller
PD	photo diode
PFGI-POF	perfluorinated polymer optical fiber
PMMA	Poly(methylmethacrylate)
POF	polymer optical fiber
PVC	Polyvinylchlorid
RIN	relative intensity noise
SBS	stimulated Brillouin scattering
SMF	singlemode fiber
SOF	silica optical fiber
SOP	state of polarization
SpBS	spontaneous Brillouin scattering
TEM	transversal electro-magnetic
VNA	vector network analyzer
VOA	variable optical attenuator

CHAPTER 1

Introduction

Practical implementation of fiber optic sensors for structural health monitoring provides crucial economic benefits regarding lower life-cycle costs of the civil infrastructure systems. Application of sensor technologies for early damage detection and prognosis leads directly to a reduction of maintenance, repair and insurance expanses. The well-established distributed fiber optic sensing methods based on scattering effects in silica fibers offer decisive advantages such as real-time remote monitoring over kilometer-long distances, operational reliability of sensors in an electromagnetic environment as well as two-dimensional sensor design using smart geosynthetics for monitoring of large geotechnical structures like dikes, railways, tunnels or embankments.

Stimulated Brillouin scattering (SBS) in silica optical fibers has been intensively studied for several decades [1–3]. The acquired knowledge has been successfully applied in fields of signal processing [4], THz signal generation [5], phase conjugation [6], lasing [7,8] and optical storage [9]. Applications for distributed strain and temperature sensing based on SBS in silica optical fibers were investigated intensively [10,11]. In all SBS sensors the Brillouin frequency shift (BFS) depends on temperature and strain. State of the art SBS sensing setups are capable of monitoring fiber lengths beyond 100 km with a few meter spatial resolution [12–14] and down to extremely short ranges with spatial resolutions ranging from centimeter [15] to millimeter [16] and micrometer [17].

Fiber optic sensors are not limited to SBS and are covering numerous fields of impacts for example temperature [18], strain [11], humidity [19–23], radiation [24,25], vibration [26], acoustics [27–29], refractive index of liquids, partial discharge in power cable joints and terminations [30], velocity of sound outside a sensing fiber [31], subsurface seismic monitoring [32], concentrations of salts in liquid compounds [33].

Compared to silica optical fibers, perfluorinated graded-index polymer optical fibers (PFGI-POFs) offer higher break down strain up to 100 % [34], minimal bending radii below 2 mm, higher sensitivity to temperature and lower sensitivity to strain [35]. These beneficial characteristics make the commercially available PFGI-POF potentiality interesting for

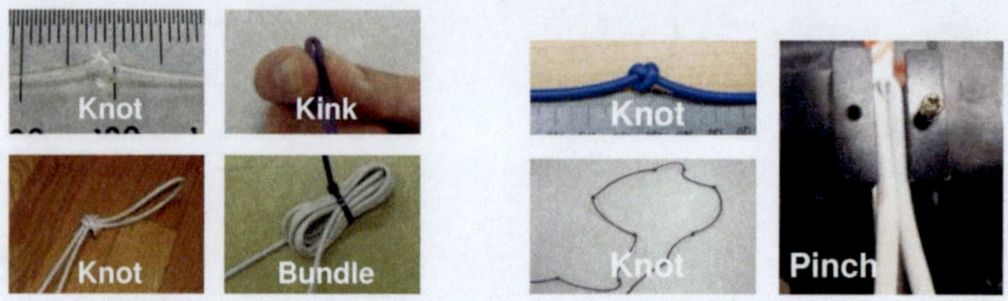

Figure 1.1: Demonstration of PFGI-POFs being knotted, bundled and pinched with and without fiber jacketing. Reprinted from `http://www.lucina.jp/eg_fontex/pdf/Tecnhical.pdf`

embedded fiber sensor applications silica fibers can not reach and withstand. Figure 1.1 underlines the fiber's flexibility under extreme mechanical influences.

Polymer optical fibers (POFs) have been employed in several areas of applications such as short-range data communication [36], polymer optical fiber amplifier [37], radiation monitoring [38], clinical monitoring of patients [39–41], cell and bacteria detection in clinical probes [42, 43], illumination [44] and humidity sensing [23].

In PFGI-POFs spontaneous [45, 46] and stimulated [47] Brillouin scattering was observed almost a decade ago. Distributed Brillouin sensing was intensively studied in the last years [48–50] and the latest results reported on spatial resolutions in the centimeter-scale with high repetition rates [51]. However, all presented distributed Brillouin fiber sensors in PFGI-POF are limited to a few tens of meters due to a high attenuation at 1550 nm ($\approx$ 250 dB/km [52]). Aiming to reduce the sensor fibers propagation loss, an alternative wavelength of 1319 nm was chosen as suggested in [18, 46]. As demonstrated in [46] Nd:YAG lasers at this wavelength are well suited to generate SBS in PFGI-POF due to a lower fiber attenuation ($\leq$ 37 dB/km at 30 °C [22]). The aim of this thesis is to design and implement a distributed Brillouin sensor based on BOFDA for PFGI-POFs. The adopted BOFDA scheme is demonstrated in a laboratory setup to achieve a sensing range of 86 m with 3.4 m spatial resolution. Furthermore, the impact of humidity changes on the Brillouin gain spectrum were empirically investigated for the first time.

This thesis deals with the advancement and implementation of distributed Brillouin sensing in PFGI-POF. A theoretical view at physical basics of Brillouin scattering is given in chapter 2. Chapter 3 introduces the light propagation in PFGI-POF and provides a comprehensive view on SBS in graded-index multimode fibers. The requirements for an reliable interconnection between PFGI-POFs and silica fibers to generate SBS are presented in chapter 4. The three major concepts of distributed Brillouin sensing are summarized in chapter 5 with special emphases on the frequency-domain approach and on distributed Brillouin sensors in PFGI-POFs. The entire relationships between SBS parameters and the environmental parameters strain, temperature and humidity are thoroughly presented in chapter 6, respectively. Chapter 7 discusses the influence of amplitude noise on the measurement accuracy of Brillouin sensors. Finally, the laboratory setup along with experimental results is laid out in chapter 8.

CHAPTER 2

Brillouin scattering in optical fibers

In general, Brillouin scattering refers to the inelastic interaction of an incident optical wave with material waves in a medium. Léon Brillouin theoretically predicted the spontaneous light scattering on thermally excited acoustic waves in 1922 [53]. In 1930 the experimental confirmation of spontaneous Brillouin scattering in liquids was given by Gross [54]. Stimulated Brillouin scattering was first observed in 1964 after the invention of the laser [55]. In the 1970s the development of low-loss glass fibers [56] resulted in intense research on optical transmission systems [57]. Brillouin scattering in optical fibers and the threshold of its stimulated process were investigated as SBS is a significant factor limiting the maximum transmittable power in optical communication systems [58, 59]. On the other hand, since then Brillouin scattering in optical fibers has been employed in various applications like high resolution spectroscopy [60], quasi-light storage and optical delay [4, 9], signal generation and processing [5, 6], sensing [10, 11], viscosity determination of liquids [61], microscopic imaging [62] and characterization of elastic properties in geosciences [63].

2.1 Spontaneous and stimulated Brillouin scattering

Light scattering is a result of inhomogeneities of the refractive index in a medium. Almost all propagation media have static fluctuations of the local dielectric permittivity causing elastic scattering of light in all directions. Thermally excited lattice vibrations in optical fibers already result in density fluctuations, which spread with the velocity of sound v_a. The scattering of light on these thermally excited acoustic waves in a medium is called spontaneous Brillouin scattering (SpBS). Due to the photoelastic effect, the sound waves induce a periodic modulation of the refractive index of the medium. The reflection and diffraction of light on (quasi)-stationary periodic gratings of refraction index changes is known as Bragg reflection [1]. The wavelength of the acoustic wave λ_a describes the spatial period of the refractive index modulation. When Braggs condition is satisfied, a wave with a vacuum wavelength of λ_P is reflected at such a periodic refractive index lattice. Equation 2.1 gives Braggs condition, where n_P is the group index of the medium for the incident pump wave and θ the angle between incident and reflected wave propagation direction. Figure 2.1 schematically illustrates the Bragg reflection.

$$\lambda_P = 2n_P\lambda_a \sin\frac{\theta}{2} \tag{2.1}$$

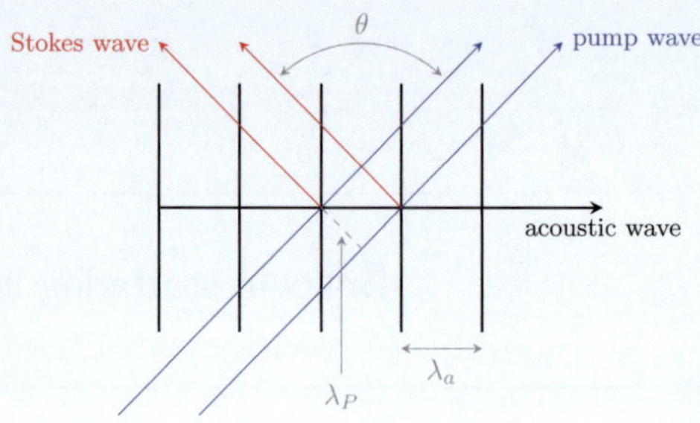

Figure 2.1:   Reflection of light on an acoustic wave. Partial reflections of the pump wave will superimpose constructively to a scattered Stokes wave if the path difference between each partial reflection is equal to λ_P.

Due to the fact that the refractive index lattice is traveling in the medium with the velocity of sound and the optical pump wave with c/n_P, the Doppler effect has to be considered for the scattered light:

$$f_S = f_P\left(1 - 2n\frac{v_a}{c}\sin\frac{\theta}{2}\right) = f_P - f_B(\theta) \tag{2.2}$$

$$f_{AS} = f_P\left(1 + 2n\frac{v_a}{c}\sin\frac{\theta}{2}\right) = f_P + f_B(\theta) \tag{2.3}$$

When the acoustic wave moves away from the pump wave, a Stokes wave with smaller frequency f_S is generated and an anti-Stokes wave of greater frequency is generated when the sound wave travels towards the pump wave. The maximum frequency offset can be observed when $\theta = \pi$ and refers to the counter propagation of the pump and Stokes wave. This maximum frequency offset is known as the material-specific Brillouin frequency shift or in short Brillouin frequency shift (BFS). In case of $\theta = \pi$ follows, that the Brillouin frequency shift f_B is equal to the acoustic frequency f_a in Eqn. 2.1.

$$f_B = 2n_P\frac{v_a}{c}f_P \tag{2.4}$$

Spontaneous Brillouin scattering on thermally excited sound waves occurs in optical fibers. For single mode fibers (SMFs) sensing systems based on SpBS have been presented [48,64,65]. However, due to arbitrary thermal motions, the SpBS propagation directions are arbitrary, too. Thus, SpBS has no dominant direction of propagation in an optically and acoustically multi-modal propagating medium and the scattered light intensity is very weak.

If the intensity of the pump wave exceeds a certain threshold at which the power transfer to the Stokes wave is higher than the attenuation it will experience, a stimulated process can occur. A small portion of SpBS Stokes waves is exactly counter propagating to the pump wave with a frequency f_S and amplified due to the Brillouin gain (see section 2.2). This process is called stimulated-amplified spontaneous Brillouin scattering, which limits the maximum transmittable power in an optical fiber. The Brillouin threshold needed to start this process will be closer described in section 2.3.

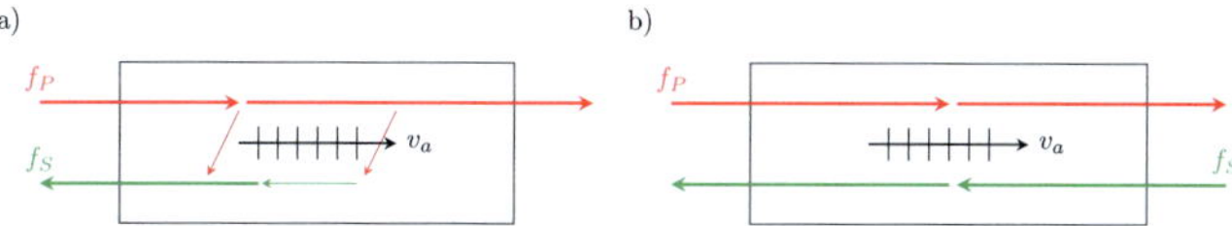

Figure 2.2: Stimulated Brillouin scattering process initiated by a) spontaneous Brillouin scattering and b) two counter propagating optical waves.

Alternately, two counter propagating optical waves at a frequency of $f_S = f_P - f_B$ can initiate stimulated Brillouin scattering (SBS) as illustrated in Fig. 2.2. The interference between the Stokes and pump waves result in an optical beating. The envelop frequency of the beating is equal to the difference frequency of both optical waves. The beating moves in the same direction as the pump wave due to the conservation of energy and momentum in the quantum mechanical model. Due to electrostriction, the intensity modulation of the beating is transfered into a density modulation of the medium. Electrostriction is associated with the property of dielectric materials and is caused by molecules which move or relocate in presence of an electric field. The density modulation of the medium is characterized by periodical variations, which associate to an acoustic wave within the medium. Consequently to the density modulations, the refractive index demonstrates the same modulations, which leads to a periodic grating in the medium fulfilling Eqn. 2.1. By increasing the pump power, the optical beating also increases, which results in raised grating structures in the medium and consequently to an exponential increment of backscattered Stokes wave power. The Stokes wave accumulates the scattered pump power along the interaction length and is therefore amplified [3]. If an optical wave at the frequency of $f_{AS} = f_P + f_B$ is injected to the counter propagating pump wave, the direction of energy flow remains as described in the process above. Thus, along the interaction length the anti-Stokes wave is attenuated depending on the pump power.

The boundary condition for optical waves is given by the Helmholtz equation [1]:

$$\nabla^2 \mathbf{E} + n^2(f)\frac{f^2}{4\pi^2 c^2}\mathbf{E} = 0 \tag{2.5}$$

Nonlinear effects in optical fibers are a result of either an intensity dependence of the refractive index or a scattering phenomena. In case of intense electromagnetic fields a homogeneous material can behave like a nonlinear medium. Consequently, the material polarization $\mathbf{P}$ is no longer proportional to the electric field $\mathbf{E}$ and can be described by a Taylor series [2]:

$$\mathbf{P} = \varepsilon_0\chi^{(1)}\mathbf{E} + \varepsilon_0\chi^{(2)}\mathbf{E}^2 + \varepsilon_0\chi^{(3)}\mathbf{E}^3 + \cdots = \mathbf{P}^L + \mathbf{P}^{NL} \tag{2.6}$$

where $\chi^{(i)}$ denotes the k-th order susceptibility and ε_0 the permittivity in vacuum. The linear susceptibility $\chi^{(1)}$ contributes dominantly to $\mathbf{P}$. The second order susceptibility $\chi^{(2)}$ causes second harmonics and sum-frequency generation. However, for media with symmetric molecule structures, like silica, $\chi^{(2)}$ vanishes. Therefore, silica optical fibers do not exhibit second order nonlinear refractive effects [66]. As written earlier, the electrostriction follows the nonlinear polarization due to a change of susceptibility and is expressed by [2]:

$$\mathbf{P}_i^{NL} = \Delta\varepsilon\mathbf{E}_i(\mathbf{r}, t) = \frac{\gamma_e}{\rho_0}\Delta\rho\mathbf{E}_i(\mathbf{r}, t). \tag{2.7}$$

$\Delta\varepsilon$ represents the change of dielectric permittivity, γ_e the electrostrictive coefficient, ρ_0 the average density of the guiding material and $\Delta\rho$ the function of the density wave in the guiding medium. Classically, the SBS is expressed by the interaction of a pump wave $\mathbf{E}_P$, a Stokes wave $\mathbf{E}_P$ and an acoustic wave $\Delta\rho$. In the following theoretical process only optical single-mode systems are considered with one acoustic wave co-propagating with the pump wave and oscillating in its own direction of propagation. The theory of optical and acoustical multi-mode systems will be discussed further in chapter 3.

$$
\begin{aligned}
\mathbf{E}_P &= \mathbf{e}_P\frac{1}{2}E_P(z, t)\,exp\left\{j(2\pi f_P t - k_P z)\right\} + c.c. \\
\mathbf{E}_S &= \mathbf{e}_S\frac{1}{2}E_S(z, t)\,exp\left\{j(2\pi f_S t + k_S z)\right\} + c.c. \\
\Delta\rho &= \frac{1}{2}A(z, t)\,exp\left\{j(2\pi f_B - k_B z)\right\} + c.c.
\end{aligned}
\tag{2.8}
$$

where $\mathbf{e}_i = \mathbf{E}_i/|\mathbf{E}_i|$ are normalized polarization vectors. The temporal and spatial propagation of Eqn. 2.8 in an electrostrictive, isotropic and homogeneous medium is described as follows [67]:

$$
\begin{aligned}
\nabla^2\mathbf{E}_P - \frac{n^2}{c^2}\frac{\partial^2\mathbf{E}_P}{\partial^2 t^2} &= \mu_0\frac{\partial^2\mathbf{P}_P^{NL}}{\partial^2 t^2} \\
\nabla^2\mathbf{E}_S + \frac{n^2}{c^2}\frac{\partial^2\mathbf{E}_S}{\partial^2 t^2} &= \mu_0\frac{\partial^2\mathbf{P}_S^{NL}}{\partial^2 t^2} \\
v_a^2\nabla^2(\Delta\rho) - \frac{\partial^2\Delta\rho}{\partial^2 t^2} - \Gamma\frac{\partial\Delta\rho}{\partial t} &= \nabla\cdot\mathbf{F}
\end{aligned}
\tag{2.9}
$$

where Γ is the acoustic damping coefficient, $\mathbf{P}_i^{NL}$ are the nonlinear polarization fields of Eqn. 2.7 and $\Delta\rho$ represents the density variations. All equations are coupled through the electrostrictive force F. This electrostrictive force is expressed using material constitutive relations [2]:

$$\nabla\cdot\mathbf{F} = -\frac{\gamma_e}{2}\nabla^2\langle E^2\rangle \tag{2.10}$$

Thus, the nonlinear polarization fields for the pump and probe wave follow as:

$$
\begin{aligned}
\mathbf{P}_P^{NL} &= \mathbf{e}_P\frac{1}{2}\frac{\gamma_e}{\rho_0}A(z, t)E_S(z, t)\,exp\left\{j(2\pi f_P t - k_P z)\right\} + c.c. \\
\mathbf{P}_S^{NL} &= \mathbf{e}_S\frac{1}{2}\frac{\gamma_e}{\rho_0}A^*(z, t)E_P(z, t)\,exp\left\{j(2\pi f_S t + k_S z)\right\} + c.c.
\end{aligned}
\tag{2.11}
$$

In case $\mathbf{P}^{NL} = 0$, the two optical waves from Eqn. 2.9 are not interacting and their powers remain constant. Therefore, their propagation is expressed by simple plain waves. If $\mathbf{P}^{NL} \neq 0$, the fields $\mathbf{E}_P$ and $\mathbf{E}_S$ are linked via the nonlinear polarization. In most cases the effect of nonlinear polarization is modest compared to the linear polarization for only a few optical wavelengths. Hence, linear solutions for the wave fields can be applied and include the slowly varying envelope approximation of [68]. Under the named assumptions the system of differential coupled equations simplifies to a set of scalar equations [67]:

$$
\begin{aligned}
\frac{\partial}{\partial z}E_P - \frac{n}{c}\frac{\partial}{\partial t}E_P + \frac{\alpha}{2}E_P &= -j\frac{k_S\gamma_e}{4\varepsilon\rho_0}AE_S \\
-\frac{\partial}{\partial z}E_S + \frac{n}{c}\frac{\partial}{\partial t}E_S - \frac{\alpha}{2}E_S &= -j\frac{k_S\gamma_e}{4\varepsilon\rho_0}A^*E_P \\
\frac{\partial}{\partial z}A + \frac{2\pi f_B - j\Gamma k_B^2}{2\pi f_B v_a}\frac{\partial}{\partial t}A &= -(\mathbf{e}_P\cdot\mathbf{e}_S)j\frac{k_B\gamma_e}{4v_a^2}E_PE_S^*
\end{aligned}
\tag{2.12}
$$

α is the optical loss of the guiding medium. During the entire interaction process, the energy and momentum have to be conserved. Therefore, the frequencies and wave vectors of all three waves are related by:

$$
\begin{aligned}
f_B &= f_P - f_S \\
\mathbf{k}_B &= \mathbf{k}_P - \mathbf{k}_S
\end{aligned}
\tag{2.13}
$$

where f_B is the frequency and k_B the wave vector of the acoustic field. According to the system of equations in (2.12), static and dynamic SBS processes are covered. For continuous wave signals ($\partial/\partial t = 0$), steady-state conditions for the acoustic field can be considered, which leads to an extinction of the time derivate term. In contrast to the low optical loss for the pump and probe wave, the acoustic wave is attenuated rapidly. In silica the free path of the phonons are usually very short ($\approx 10^{-6}$ m). Hence, the amplitude of the acoustic wave is expressed by [2]:

$$
A = j\frac{k_B\gamma_e}{2v_a\Delta f_B}(\mathbf{e}_P\cdot\mathbf{e}_S)E_PE_S^*
\tag{2.14}
$$

where $\Delta f_B = \Gamma k_B^2$ is the bandwidth of the Brillouin scattering. One observes, that the Brillouin linewidth is directly scaling with the acoustic damping coefficient, which represents the average lifetime of a phonon within the guiding medium [1]. Eventually, inserting Eqn. 2.14 into Eqn. 2.12 delivers two coupled equations, describing the relation between pump and Stokes wave in the SBS process:

$$
\begin{aligned}
\frac{\partial}{\partial z}E_P &= -(\mathbf{e}_P\cdot\mathbf{e}_S)\frac{k_Bk_P\gamma_e^2}{8\varepsilon\rho_0 v_a\Delta f_B}\frac{E_P|E_S|^2}{1-j(2\Delta\nu/\Delta f_B)} - \frac{\alpha}{2}E_P \\
-\frac{\partial}{\partial z}E_S &= -(\mathbf{e}_P\cdot\mathbf{e}_S)\frac{k_Bk_S\gamma_e^2}{8\varepsilon\rho_0 v_a\Delta f_B}\frac{E_S|E_P|^2}{1-j(2\Delta\nu/\Delta f_B)} - \frac{\alpha}{2}E_S
\end{aligned}
\tag{2.15}
$$

where $\Delta\nu$ represents the detuning frequency ($\nu - f_B$) between a current frequency state ν and the Brillouin frequency shift given in Eqn. 2.1. The real parts of Eqn. 2.15 cause an energy transfer and consequently an optical gain or an optical loss. Whereas the imaginary parts lead to a nonlinear phase shift. Considering steady state conditions for continuous pump and probe waves, the equations can be further simplified as the lifetime of the acoustic wave is negligibly small compared to the lifetime of the pump wave. Furthermore, the introduction of optical intensities $I_i = \frac{1}{2}n\varepsilon_0 c|E_i|^2$ for pump and Stokes wave enables Eqn. 2.15 to be written as [1,3]:

$$\frac{\partial I_P}{\partial z} = -g_B(\Delta\nu)I_P I_S - \alpha I_P$$
$$\frac{\partial I_S}{\partial z} = -g_B(\Delta\nu)I_P I_S + \alpha I_P. \qquad (2.16)$$

The new parameter g_B is introduced to describe the Brillouin gain coefficient and its dependency of frequency detuning. The loss resonance f_{AS} caused by SBS is mathematically given in the same manner as for the Stokes scattering. Only the sign of the Brillouin gain coefficient is changed and $f_{AS} > f_P$. The spectral response of the Brillouin loss is equal to its gain as illustrated in Fig. 2.3.

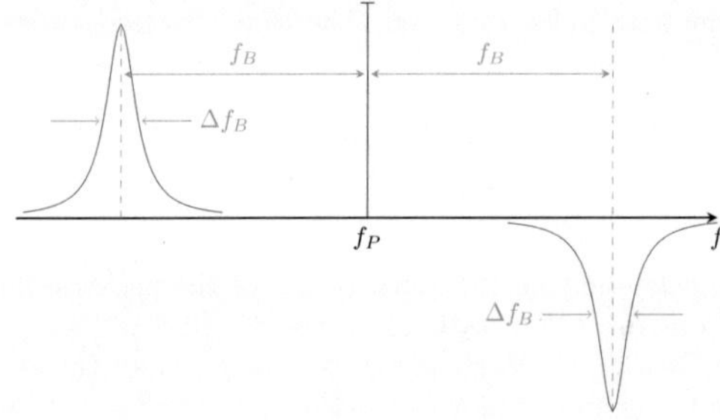

Figure 2.3:   Schematic illustration of the Brillouin gain and loss resonances with respect to the pump wave.

Assuming the pump wave power is not exceeding the Brillouin threshold to cause stimulated-amplified spontaneous Brillouin scattering, the intensity of the pump wave is only affected by the attenuation of the medium [3]:

$$I_P(z) = I_P(z = 0)\, exp\{-\alpha z\} \qquad (2.17)$$

where z is introduced as a spatial parameter along the guiding medium. $z = 0$ refers to the beginning of the medium at which the pump wave is launched. $z = L$ to the end of the medium at which the Stokes wave is launched. The total length of the medium is L. Thus, the pump intensity at the position L is:

$$I_P(z = L) = I_P(z = 0)\int_0^L exp\{-\alpha z\}\, dz = \frac{I_P(z = 0)}{\alpha}(1 - exp\{-\alpha L\}) = I_P(z = 0)L_{eff}.$$
$$(2.18)$$

L_{eff} represents the effective length of the medium. It is note, that the SBS does not only depend on intensities but also on the interaction length at which the power density remains sufficiently high [3]. The attenuation of the guiding medium and the total fiber length are relevant factors to describe the effective length of nonlinear interaction. Consequently, the Stokes wave in Eqn. 2.16 can be written as:

$$\frac{\partial I_S}{\partial z} = I_S\left(-g_B I_P(z) + \alpha\right)$$

(2.19)

Combining Eqn. 2.19 and 2.18 gives the intensity of the Stokes wave at the Position $z = L$:

$$I_S(z = L) = I_S(z = 0)\ exp\left\{\frac{-g_B P_P(z=0)L_{eff}}{A_{eff} + \alpha L}\right\}.$$

(2.20)

Hence, the intensity of the Stokes wave at the beginning of the medium is [58]:

$$I_S(z = 0) = I_S(z = L)\ exp\left\{\frac{g_B P_P(z=0)L_{eff}}{A_{eff} - \alpha L}\right\}.$$

(2.21)

Note, that the intensity of the pump wave is given as the ratio of power P_P and effective illuminated area in the medium A_{eff}. Therefore, next to the intensities of the optical waves and the length of optical interaction also the effective illuminated area inside the medium influences affects the SBS process [1–3]. Furthermore, for pump and Stokes wave powers above the Brillouin threshold Eqn. 2.21 is not valid and does not consider the polarization dependence between pump and Stokes wave during the SBS process.

2.2 The Brillouin gain spectrum

The Brillouin gain coeffcient $g_B(\Delta\nu)$ and its spectral distribution is approximated by a Lorentzian function. This approximation originates from the acoustic wave in the SBS process as it can be considered a force and damped oscillation [1, 2].

$$g_B(\Delta\nu) = g_p\frac{(\Delta f_B/2)^2}{(\Delta\nu)^2 + (\Delta f_B/2)^2}$$

(2.22)

where g_P refers to the peak value of the Brillouin gain coefficient. Its definition was indirectly given in Eqn. 2.15 [2]:

$$g_p = \frac{\eta_P 2\pi n^7 p_{12}^2}{c\lambda_P^2 \rho v_a \Delta f_B}$$

(2.23)

with λ_P denoting the pump wavelength and p_{12} referring to the longitudinal elasto-optic coefficient. η_P indicates a dependency on the polarization of the SBS interaction. For optical fibers the state of polarization is not constant due to local birefringence that statistically varies along the fiber in strength and orientation [66]. Thus, the inference of pump and stokes wave causing the beating signal in an optical fiber is dependent on the state of polarization between pump and Stokes wave to each other. η_P represents the polarization mixing efficiency of the pump and Stokes wave given by $\eta_P = |\mathbf{e}_P \cdot \mathbf{e}_S|^2$. Furthermore, a strong dependency on the refractive index with a power to 7 underlines the influence of material properties. A normalized gain distribution of the Brillouin gain spectrum (BGS) is plotted in Fig. 2.4. As mentioned earlier, the amplification of the

Stokes wave comes with a nonlinear phase change. Thanks to the Kramers-Kronig relation, the real and imaginary parts of the susceptibility of a medium are connected and links the absorption coefficient to the refractive index in a dispersive medium. The normalized imaginary part of the Brillouin gain, which corresponds to the phase change during the SBS amplification, is plotted in Fig. 2.4, too.

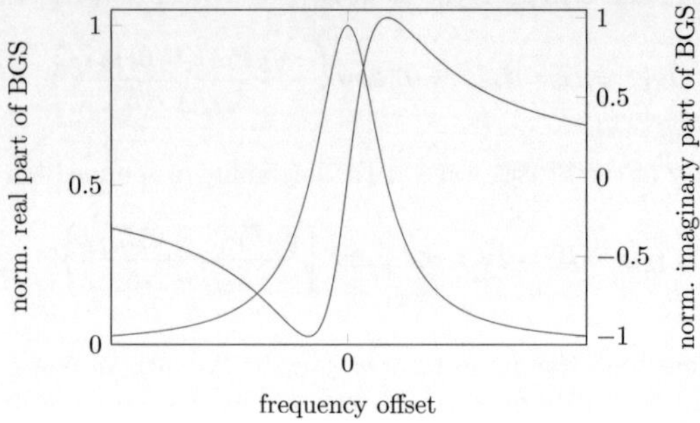

Figure 2.4: Normalized Brillouin gain spectrum splited in real and imaginary part.

However, if pump pulses are temporally shorter than the phonon life time or the linewidth of the pump source, the corresponding pump pulse spectrum is broader than the intrinsic Brillouin linewidth. Consequently, the resulting Brillouin gain spectrum is deformed and Eqn. 2.23 is need to be enhanced to [3]:

$$g_p = \frac{\eta_P 2\pi n^7 p_{12}^2}{c\lambda_P^2 \rho v_a \Delta f_B} \left(\frac{\Delta f_B}{\Delta f_B * \Delta f_P} \right) \tag{2.24}$$

where Δf_P denotes the linewidth of the pump source and $*$ the convolution operation. In case $\Delta f_P \ll \Delta f_B$ the effect of the pump source is negligibly small and the Brillouin linewidth is rather affected by material properties. Thus, narrow linewidth laser sources are in preferential use for the evaluation of a Brillouin gain spectrum.

Among the Brilllouin gain coefficient and the Brillouin linewidth, the Brillouin frequency shift characterizes the gain spectrum. Eqn. 2.4 denotes the dependencies of the frequency shift by the refractive index, the wavelength of the pump wave and the velocity of sound within the medium. The presented derivation of the SBS process assumed one acoustic wave propagating along the direction of the pump wave. In optical fibers this acoustic wave corresponds to a longitudinal wave oscillating in a radial direction [1,58]. Thus, the velocity of sound for this specific mode in cylindrical media is expressed by $v_a = \sqrt{\epsilon/\rho_0}$, where ϵ is the elastic modulus and ρ_0 the average density of the guiding medium [1]. Consequently, Eqn. 2.4 enhances to:

$$f_B = \frac{2n \, v_a}{\lambda_P} = \frac{2n \sqrt{\epsilon/\rho_0}}{\lambda_P} \tag{2.25}$$

Therefore, changes in the medium material parameters n, ϵ and ρ are causing Brillouin frequency shifts. Environmental conditions like temperature and tensile strain changes along an optical fiber relate to specific Brillouin frequency shift changes. For silica optical fibers the frequency change demonstrates a linear relation to temperature and strain changes. The slope values of the linear relation are forming the basis for Brillouin sensing in optical fibers and were determined for temperature and strain changes to be $C_T = \partial BFS/\partial T \approx 1$ MHz/K and $C_S = \partial BFS/\partial S \approx 500$ MHz/%, respectively [11].

2.3 Brillouin threshold

An important parameter of Brillouin scattering is the threshold of pump power at which a significant part of the power is reflected. Several definitions of the threshold in optical fibers are given in the literature [3]:

- The input pump power is equal to the backscattered Stokes wave power [59].

- the input pump power at which the power of the backscattered Stokes wave is equal to the transmitted power [69].

- The input pump power that is required for a strong increase of the Stokes power or a strong decay of the pump power [70].

- The input pump power at which the backscattered power at the fiber input corresponds to 1 % of the pump input power [71].

In most practical applications the threshold is defined as the input pump power at which the backscattered and transmitted powers are equal [69]. Assuming the pump wave source has a significantly smaller linewidth than the Brillouin linewidth of the guiding medium, the Brillouin threshold P_{th} can be approximated by:

$$P_{th} = 21 \frac{K_B A_{eff}}{g_p L_{eff}} \tag{2.26}$$

where K_B is a polarization factor indirectly scaling to the polarization mixing efficiency $K_B = 1/\eta_P$. Polarization maintaining fibers will take a value of $K_B = 1$, whereas randomly polarized waves are represented by a value of $K_B = 1.5$ [72]. g_p corresponds to the maximum value of the Brillouin gain and L_{eff} is the effective length of the medium.

$$L_{eff} = \frac{1 - exp\left\{-\alpha L\right\}}{\alpha} \tag{2.27}$$

where α is the attenuation coefficient and L the total length of the fiber. Due to the rapid development of optical fibers the assumed fiber loss of 2 dB/km, originating back to the 1970s, Eqn. 2.26 needed to be adopted for modern silica fibers by correcting the factor 21 to 19 by the European Cooperation for Scientific research [73]:

$$P_{th} = 19 \frac{K_B A_{eff}}{g_p L_{eff}} \tag{2.28}$$

CHAPTER 3

Propagation of light in POF

The majority of fiber optic sensors use SMFs as a sensing medium. Beneficial characteristics such as very low optical loss, the absence of multi-modal behavior (launch conditions, modal dispersion and mode coupling) and high availability of numerous commercial fibers are reasons for their use. However, the emergence of polymer optical fibers (POFs) over the last years offers several advantages and new opportunities in sensing. POFs are cheap, easily manageable concerning assembly of cables and, especially, their connectors (which is due to the large core diameters and the resulting high numerical aperture). POFs are mechanically robust, which designates them as attractive an attractive tool for field applications in rough conditions. Finally, they are able to manage strain which leads to fiber elongations of more than 100 % [34]. The strain handling forms an enormous advantage over silica fibers with their maximum strain of 1-2 % [74].

In this chapter POFs are introduced with special emphasis on perfluorinated graded-index POF (PFGI-POF), that will be utilized for SBS generation. Differences in multimode propagation and mode coupling mechanisms between PFGI-POF and silica optical fibers are outlined. The dependence of SBS on optical modes in multimode fibers is presented. Based on these considerations and on the gained information from chapter 2, theoretical assumptions for SBS sensing in PFGI-POF are given. This chapter aims on the understanding of light propagation and SBS interaction in PFGI-POF as prerequisite for the later measurements.

3.1 Perfluorinated graded-index POF

Before the emerge of PFGI-POF, poly(methylmethacrylate) (PMMA) was the standard core material for POFs. The most common PMMA-POF has a step-index profile with a core diameter of 1 mm [36], but grad-index PMMA-POFs with 675 μm core and multicore PMMA-POFs are commercially available, too [75]. Due to their large core diameters, the resulting high numerical apertures enable an easy assembly of fibers and connectors. Thus, PMMA-POFs have been applied in several areas such as short-range data communication [36], polymer optical fiber amplifier [37], radiation monitoring [38], clinical monitoring of patients [39–41], cell and bacteria detection in clinical probes [42,43], illumination [44] and humidity sensing [23]. The attenuation of PMMA-POFs in the visible spectral range of light largely originates from the overtone vibration absorption of the C-H bonds in the polymer [76] since each monomer in a PMMA consists of 8 C-H bonds as shown in Fig. 3.1a).

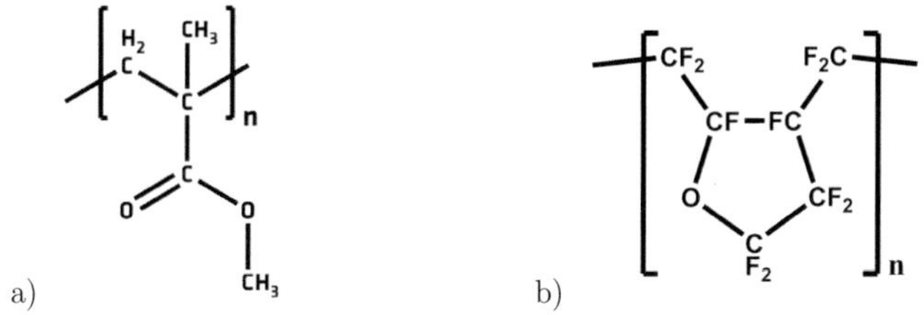

Figure 3.1: Molecular structure of a) PMMA and b) CYTOP.

This strong vibrational attenuation drastically limits the spatial range of a POF transmission link or a sensor. The fiber attenuation was considerably reduced by replacing hydrogen with heavy hydrogen. However, these deuterated POF are sensitive to water vapor absorption which leads to an additional increase of fiber attenuation [52]. The replacement of hydrogen with heavier fluorine atoms shifted the C-F vibration bands away from the spectra range used in telecommunication (850 nm to 1550 nm) [52]. Among other fluorinated polymers, the lowest absorption in an optical fiber was obtained using the cyclic transparent optical polymer (CYTOP) poly(perfluorobutenylvinylether), which is illustrated in Fig. 3.1b). In 1997, the first graded-index POF based on CYTOP was reported with an attenuation of 50 dB/km at a wavelength of 1300 nm [77]. Since then, the manufacturing process of CYTOP POF (equal to PFGI-POF) has been undergone intense research [52,78,79]. Nowadays, the attenuation of approximately 1040 nm has been reduced to roughly 10 dB/km [52]. CYTOPs theoretical lowest attenuation was calculated to be 0.3 dB/km at a wavelength of 1300 nm, which is comparable to silica optical fibers. Figure 3.2 shows the comparison of fiber attenuation as a function of wavelength for PMMA and CYTOP.

However, this theoretical attenuation minimum has not been reached, yet. Currently, limitations are largely attributed to the manufacturing process and material contamination. This is due to impurities and material crystallization, which form statistically distributed scattering centers along the fiber [80]. The water absorption ratio by weight of pure CYTOP is measured to be <0.01 % [52] and is far below that of PMMA (up to 2.7 %, depending on the level of residual monomer in the PMMA resin [81]). Though, temperature and humidity affect the optical attenuation at specific wavelength regions as shown in chapter 6.

Nowadays PFGI-POFs are commercially available in different configurations. Core diameters of 50 μm, 62.5 μm and 120 μm were specifically designed to facilitate interconnections between PFGI-POFs and their silica equivalents. The refractive index change across the graded-index profile was designed to match with the silica equivalents to obtain comparable numerical apertures (NA = 0.19 ± 0.015 [82] and NA = 0.2 [83] for PFGI-POF and silica

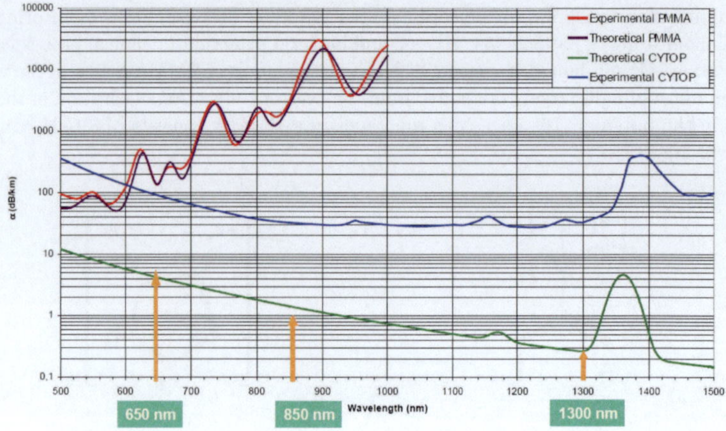

Figure 3.2: Spectral attenuation of PMMA and CYTOP POFs. Reprinted from [80]

GI-MMF, respectively). The focus in this thesis is on the 50 μm PFGI-POF, which is schematically compared to its silica equivalent and to a SMF in Fig. 3.3. It is worth noting, that the over-cladding material of the PFGI-POF predominantly consists of polycarbonate and PFGI-POFs equipped with an additional fiber jacketing. However, they are not investigated in this thesis.

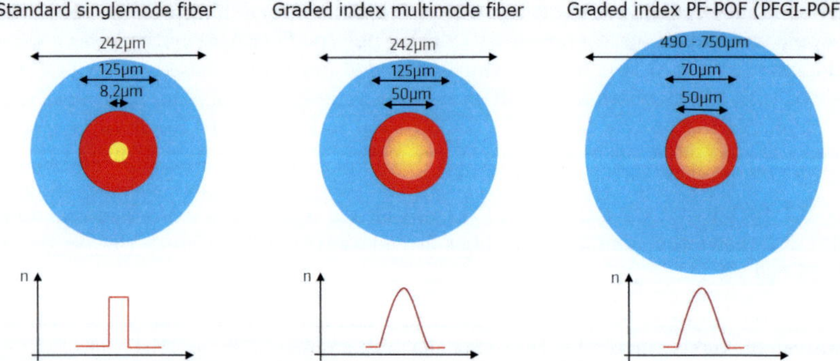

Figure 3.3: Geometry of standard SMF, graded-index multimode fiber and PFGI-POF with their corresponding schematic refractive index profiles across the fiber facets. Not drawn to scale. Yellow represents the fiber core, red the fiber cadding and blue the fiber over-cladding.

The graded-index profile in PFGI-POF is achieved by an interfacial gel polymerization method [52]. The mixture is composed of two kinds of monomers (the core host material CYTOP and a dopant) and they are inserted into a tube with a diameter equivalent to the preform diameter. The mixture is heated up to 80 °C to be preliminary liquefied. During the heating process, the formation of a gel layer is done in the inner wall of the tube (polymer gel phase) and the smaller size monomer diffuse from the edge of the preform to the center in order to form the graded-index profile. The graded-index profile is correlated to the dopant distribution inside the preform [52, 80]. The process of thermal-induced dopant diffusion is schematically illustrated in Fig. 3.4.

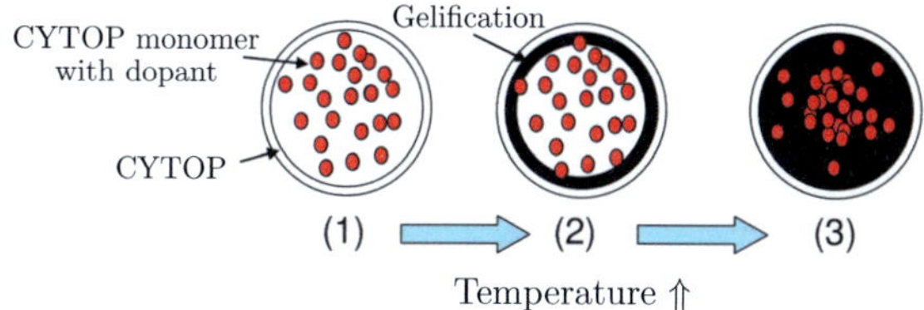

Figure 3.4: Interfacial gel polymerization method used to manufacture a graded-index profile in PFGI-POF.

3.2 Propagation of light in multimode fibers

In general, dispersion is the limiting factor in low budget optical communication systems concerning the maximum bandwidth. The broadening of optical light pulses traveling along the a fiber also affects the performance of FOSs. Dispersion leads to a degradation of spatial resolutions and distance accuracy with increasing fiber lengths. The most relevant sources of dispersion are chromatic dispersion and modal dispersion, which are both considered in this section.

The chromatic dispersion of bulk CYTOP has been reported to be around -54 ps/(nm km) and -9 ps/(nm km) at a wavelength of 845 nm and 1300 nm, respectively [84]. Figure 3.5 shows the wavelength dependency of material dispersion for silica, PMMA and CYTOP calculated from the refractive indexes.

An observation can be made, that the material-induced chromatic dispersion of CYTOP demonstrates smaller values compared to silica. The total chromatic dispersion of a PFGI-POF was measured with a value of -88 ps/(nm km) at a wavelength of 845 nm and is expected to be lower around 1300 nm [85]. The zero dispersion wavelength of the employed PFGI-POF for later experiments is specified for a wide spectral range between 1200 nm and 1650 nm [82]. However, exact information on zero dispersion wavelength and chromatic dispersion could not be obtained from the manufacturer. Nevertheless, the chromatic dispersion of the PFGI-POF is assumed to be negligibly small for wavelengths around 1300 nm since sensing applications in this fiber are limited to a few hundred meters. Consequently, chromatic dispersion would have a negligible small influence on the measurement results and is therefore not further considered.

In contrast to SMF, where the light solely propagates in the fundamental mode, multimode fibers populate up to thousands of optical modes. Therefore, the bandwidth in multimode fibers is largely limited by the modal dispersion originating from different group velocities of the propagating modes. These different mode group velocities are causing a consecutive times of arrival at the photo detector. Thus, the effect is called differential mode delay

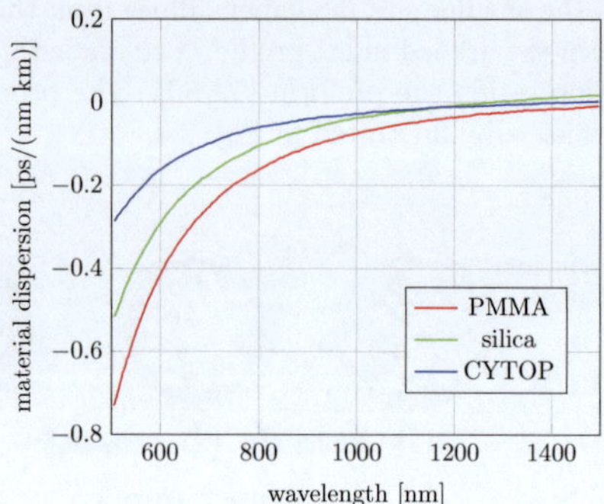

Figure 3.5: Comparison of material-induced chromatic dispersion as a function of wavelength for silica, PMMA and CYTOP. The plotted data were extracted
from [52].

(DMD). Even though the optimized graded-index profile of the PFGI-POF is designed
to minimize the delay changes between excited optical modes, the DMD still might be a
bandwidth-limiting factor as the production of the graded-index profile varies randomly
along the fiber due to its uncontrolled character.

The most striking difference between silica and polymer optical fibers are the mode coupling
mechanisms. Graded-index silica fibers exhibit only a very small power exchange between
different mode groups. This behavior is largely attributed to their high purity and high
length scale of the refractive index induced during the drawing process [86]. The stable
propagation within each excited mode results in a linear increase of pulse broadening with
increasing fiber length. Extrinsic mechanical influences like fiber bends can severely alter
the DMD by introducing intergroup coupling into higher order modes or even decoupling
of higher order mode groups. Thus, the launch condition of the light source into the fiber
(overfilled or underfilled) and the mechanical impacts along the fiber influence the transfer
function of the fiber.

Mode coupling is generally stronger in POF, irrespective their core materials and index
profiles, compared to their silica equivalents. Therefore, mode coupling is the crucial
impact on the bandwidth of a POF. Even though mode coupling has been determined to
be the decisive factor in PMMA-POFs [87,88] and PFGI-POFs [85,89,90], most theoretical
POF bandwidth calculations did not consider the impact of power coupling between
different mode groups. For graded-index PMMA-POFs strong mode coupling has been
identified as the reason to increase the bandwidth exceeding the theoretical value by far
(3 GHz to 0.43 GHz for 100 m graded-index PMMA-POF [87]). The reason is given in the
strong coupling of energy between guided mode groups on a small length scale. Random
redistribution of energy between slower and faster mode groups along the fiber caused
reduced delay differences compared to fibers without mode coupling. Rather than a linear
dependency of pulse broadening as known from silica fibers, the pulse broadening due to
mode coupling Ξ on a fiber length L is expressed as:

$$\Xi \propto L^{\varsigma} \tag{3.1}$$

with $\zeta = 1$ for no present mode coupling. The pulse broadening reduces to an almost square root dependency on L to $\zeta = 0.5$. Consequently, the fiber bandwidth B in presence and absence of mode coupling in were described by [87]:

$$\sqrt{\frac{L_o}{L_c}} = \frac{B_c}{B_o} \tag{3.2}$$

where the indexes c and o correspond to the presence and absence of mode coupling, respectively. The coupling length of a graded-index PMMA-POF has first experimentally been determined in [87] with $L_c = 2.1$ m and $\zeta = 0.6$. The coupling length of a step-index PMMA-POF was determined to be 15 m [88].

Various intrinsic and extrinsic origins lead to mode coupling in PFGI-POF. Intrinsic mode coupling is a result of fluctuations in density and concentration [88]. During the drawing process, thermally excited fluctuations of density and composition as well as polymer orientation were "frozen" into the fiber during the cooling period from the melt [91]. This results in microscopic heterogeneities scaling up to the micrometer range causing intermodal coupling [78, 89]. A post-fabrication fiber annealing at a constant humidity level demonstrated a molecular alignment relaxation leading to more stabilized conditions favorable for sensor applications in POF based on PMMA [92] and CYTOP [93]. The intrinsic attenuation measured on bulk samples of CYTOP demonstrated to vary from 5 dB/km for undoped samples up to 10 dB/km for doped samples [89]. Generally, the degree of mode coupling and the absolute attenuation of the PFGI-POF are predominantly determined by extrinsic effects and perturbations. Extrinsic sources like voids, cracks, mircobends, index profile variations and diameter variations formed during the process of fiber drawing are considered to have the dominant impact on mode coupling [91]. Improved drawing processes reduced the impurity contamination of more recent fiber batches but relatively strong scattering centers can still be observed irregularly distributed along the fiber [94].

The spatially short coupling lengths and the intergroup coupling have been identified as the main mechanisms leading to the high bandwidth of the PFGI-POF. Literature reports on 40 Gb/s over 100 m and beyond [85, 90, 95]. Any launch condition up to a radial offset of 25 μm and the overfilled launch condition into a 200 m long PFGI-POF, resulted in equivalent transmission performances of a 40 Gb/s link, which indicates complete mode coupling [90]. In addition, incoherent optical frequency-domain reflectometry (iOFDR) measurements for different modal excitation did not exhibit significant deviations on the Rayleigh backscatter traces [94]. Simulation results reported on mode coupling as the predominant factor for the low DMD in PFGI-POF and the differential mode attenuation has negligible influence [89, 90].

Due to mode coupling, the PFGI-POF offers beneficial characteristics for distributed backscatter sensing techniques. The strong mode coupling reduces DMD and makes the PFGI-POF relatively insensitive to changes of the modal excitation. The decoupling of mode groups mechanical impacts or other extrinsic impacts are almost immediately compensated for on a small length scale. Therefore, precise fiber length and attenuation changes can be measured ensuring relative insensitivity to mechanical impacts on the PFGI-POF as sensor fiber. The relatively strong backscatter fluctuations and the presence of scattering centers along the fiber considerably improve distributed length change measurements [94]. However, the impact of these scattering centers on SBS remains unknown and are subject of this thesis (see chapter 8).

3.3 Brillouin scattering in PFGI-POF

In chapter 2 the Brillouin scattering was introduced for three interacting waves - the optical pump and Stokes waves as well as one acoustic wave, co-propagating in direction of the pump wave in a Brillouin gain configuration. In single mode fibers, neglecting birefringence, there is only one effective optical refractive index n, while each acoustic mode supported by the fiber causes a different BFS due to different acoustic velocities. Thus, the overall BGS is a superposition of various Lorentzian spectral shapes. Each Lorentzian is characterized by a different BFS and weighted by the acousto-optic effective area [96] evaluating from the spatial overlap between acoustic mode and optical mode field distributions. Spatially resolved multiple acoustic modes in SMFs measured by distributed Brillouin sensing have been reported in [97]. However, the cross-evaluation of different acoustical modes does not offer further possibilities of distinction between temperature and strain influences on the SMF [97].

In multimode fibers the process of Brillouin scattering becomes increasingly complex as generally many optical and acoustical modes are involved. Each pair of optical modes can interact through an acoustical mode, which travels in the same direction of the pump. These interactions contribute to the overall BGS and are weighted depending on the overlap integral between the three involved modes [98]. The interaction of two identical optical modes is known as intramodal SBS and the interaction between different optical mode is called intermodal SBS [99].

The SBS in few-mode fiber has been numerically [98, 99] and experimentally [100–102] studied. The results demonstrate a BGS dependence on the involved optical modes. In addition, the intermodal SBS gain is reported to be of the same order of magnitude or larger [99, 102] than the intramodal SBS gain.

The majority of the literature on SBS in graded-index multimode fibers focuses on the evaluation of the SBS threshold [103–105]. The Brillouin gain coefficient in multimode fibers is reported to be a function of radial distance to the fiber core [103]. In case of overfilled launch conditions, Eqn. 2.26 has been experientially confirmed for a variation of fiber core diameters [104]. For underfilled launch conditions two main adjustments have to be considered. An average numerical aperture is introduced to the gain equation and the optimum Gaussian radius of the corresponding mode field needs to be considered for the calculation of the effective illuminated area [105]. It was also found, that the gain coefficient for lower-order optical modes significantly exceeds the gain of higher-order modes [106].

Generally, the Brillouin gain spectrum in graded-index multimode fibers is expressed as the sum of the spectral density distributions from each triad of modes [107]:

$$BGS_{P,S}(\Delta\nu) = \sum_{a=1}^{N_{ac}} g_{P,S,a} \frac{1}{1 + \left(\frac{\Delta\nu - BFS_{P,S,a}}{\Delta f_B/2}\right)^2} \tag{3.3}$$

where P, S and a are referring to the mode indexes of the pump, Stokes and acoustic wave, respectively. N_{ac} is the number of acoustic modes. In order to calculate the BFS for higher different acoustical mode, the following distribution of longitudinal velocities is assumed for silica fibers [96]:

$$v_a(r,\theta) = v_{10}\left(1 - \frac{7.8(n(r,\theta) - n_{cl})}{n_{cl}}\right) \tag{3.4}$$

with $v_{10} = 5944$ m/s as the first appearing longitudinal acoustic mode velocity into the cladding for silica optical fibers. n_{cl} is the refractive index of the fiber cladding. By assuming a propagation constant for the acoustic wave equal to the sum of propagation constants of the two interacting optical modes, the BFS associated for each acoustic mode can be determined [108]. For the sake of simplification, the BFS of acoustic modes can be determined by imposing the phase-matching condition between the acoustic mode and the propagating constant of the fundamental optical mode (index = 1) [108]. Thus, the BFS corresponding to a specific mode triad can be computed by applying the following relation [102]:

$$BFS_{P,S,a} = BFS_{1,1,a}\frac{n_{eff,P} + n_{eff,S}}{2n_{eff,1}} \tag{3.5}$$

where $n_{eff,i}$ is the effective refractive index of the i-th optical mode. The Brillouin gain graded-index multimode fiber is given by [98]:

$$g_{P,S,a} = \frac{c\mu_0\gamma_e^2\Delta f_B}{2\lambda_P\rho_0\epsilon_0 n^2\ BFS_{P,S,a}}\left(\iint \Delta\rho_a(r,\theta)E_P(r,\theta)E_S(r,\theta)\ drd\theta\right)^2 \tag{3.6}$$

where μ_0 is the magnetic permeability in vacuum. The integral corresponds to the fiber cross section involved in the SBS process. In case of intramodal SBS the fiber cross section Eqn. 3.6 reduces to the inverse of the acousto-optic effective area [96].

Due to the cylindrical design of the graded-index fibers, the interaction between a couple of optical modes occurs through two acoustic family modes. Thus, the acoustic modes involved in SBS also have an azimuthal component. Based on the Eqn. 3.3 - Eqn. 3.6 a finite-element analysis for graded-index silica multimode fibers (core diameter of 50 μm) was presented in [108]. Figure 3.6 displays how the numerical decomposition of pump and Stokes fields coincide with the fundamental mode of fiber.

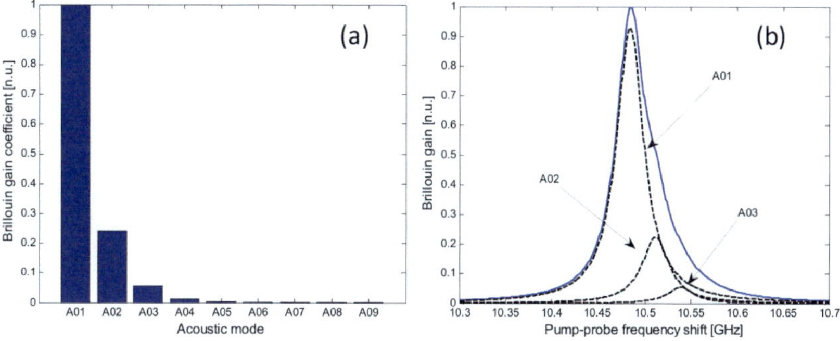

Figure 3.6: Normalized Brillouin coefficients (a) and decomposed Brillouin gain spectrum (b) computed for an optical fundamental mode to fundamental mode pair. Reprinted from [108].

The acoustic mode indexes are given as azimuthal, radial. The Brillouin scattering from
higher-order acoustical modes is responsible for an asymmetric shape of the overall BGS,
broadening towards higher frequencies. The spectral distance between adjacent resonances
(≈ 27 MHz) is smaller than the linewidth of each resonance. Therefore, the various
contributions accumulate to a linewidth of the overall BGS equal to 45 MHz.

Furthermore, the SBS interaction between an optical fundamental mode and a second-order
optical mode was numerically determined. The results demonstrated that an azimuthal
mode oscillates with several overtones in radial direction. The reason for the azimuthal
acoustic mode is seen in the anti-symmetric field distribution of the second-order optical
mode. In this scenario the Brillouin gain rapidly decreases with increasing radial mode
numbers. The contribution from higher-order acoustical modes (greater A17 and A1-A10)
can be considered negligible. The spectral distance between adjacent acoustic resonances
remains ≈ 27 MHz, though the resulting overall BGS is broadened (linewidth of 58 MHz)
in virtue to the larger number of gain associated high-order acoustic modes.

The interaction of a second-order optical mode pair is shown in Fig. 3.7. In contrast to the
other two scenarios, 17 dominant acoustical modes were found to contribute to the overall
BGS split into two predominant acoustic azimuthal mode families. Due to the number of
involved acoustic modes, the overall BGS widens to a linewidth of 81 MHz.

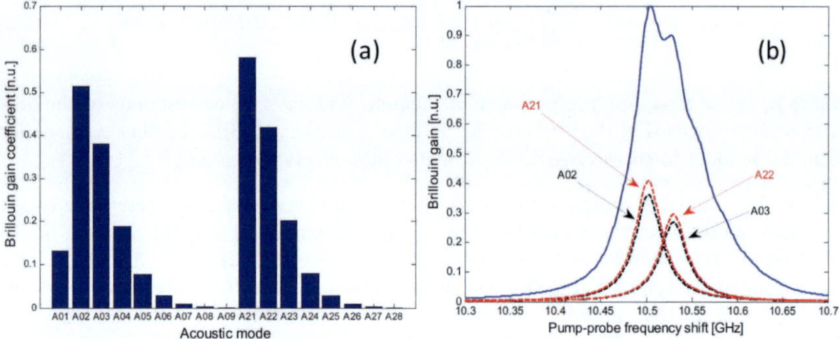

Figure 3.7: Brillouin gain coefficients (a) and Brillouin gain spectrum (b) computed for a
second order optical mode pair. Reprinted from [108].

All these results obtained from a finite-element analysis have been confirmed by experimental measurements of the BGS for given misalignments. Due to the lack of mode coupling
in silica graded-index fibers, a distributed measurement of a multimode fiber underlines
the spectral asymmetric shapes of the BGS for a predominantly optical fundamental mode
pair. Figure 3.8 shows the measured BGS at different spatial fiber segments and the
corresponding backscattering intensities assumed for two dominant acoustic modes. The
second dominant acoustic mode was found in the red marked spectral range.

It can be observed, that the distortion of the overall BGS remains along the entire fiber
length. Furthermore, the overall BGS demonstrates a slight frequency shift of its main peak
towards smaller frequencies (a detailed view is given in Fig. 3.9). This frequency shift can
be associated with the fiber loss, which attenuates additional spectral contributions caused
by higher-order acoustic modes. The backscattering intensities for those backscattered

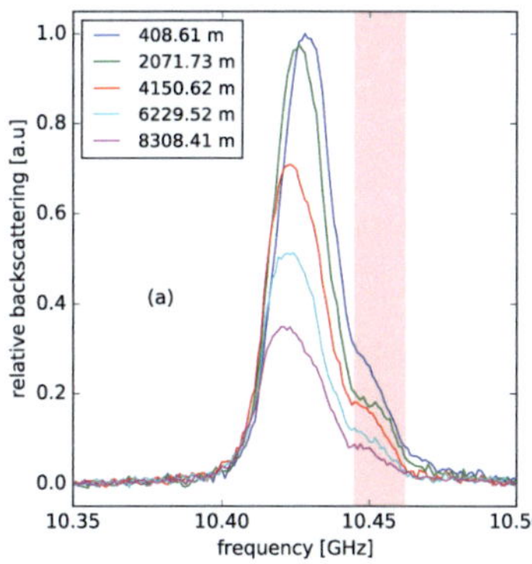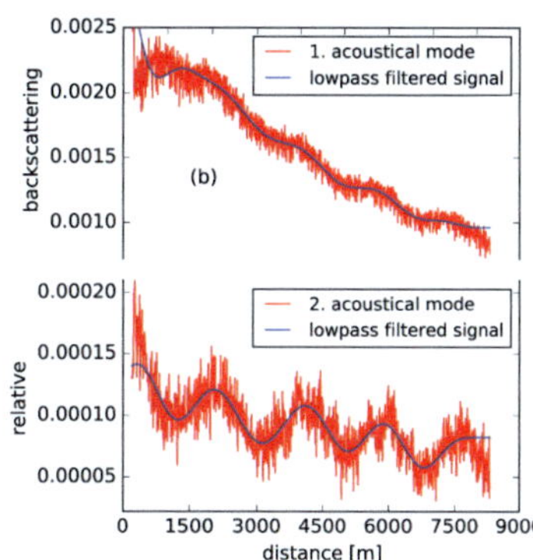

Figure 3.8: Distributed SBS measurement in a 8 km long silica graded-inded multimode
fiber with a core diameter of 50 μm. (a) Selection of spectral responses at
given spatial fiber segments. (b) Evaluated backscattering intensities for both
dominant acoustic modes, respectively.

dominant modes underline the impact of the fiber loss. The observed spatial oscillation of
backscattering intensities corresponds to the linewidth of the employed laser source.

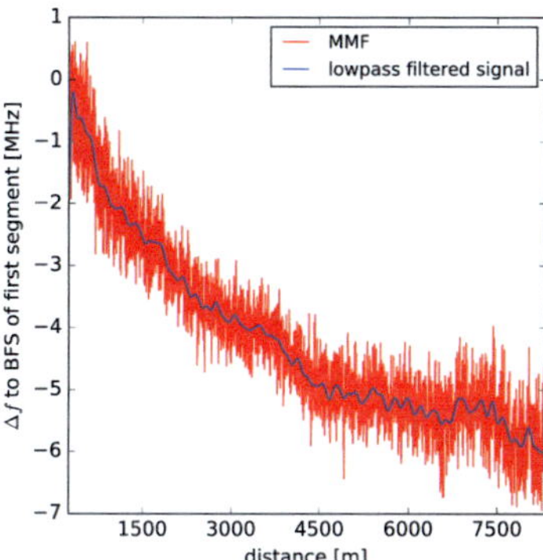

Figure 3.9: Determined BFSs of the overall peak along the fiber length. The BFSs are
given as a difference to the BFS of the first fiber segment.

For the sake of clarification, Eqn. 3.3 - Eqn. 3.6 assume no mode coupling along the fiber
and the modal field distributions of pump and probe wave are considered uniform through
the whole fiber length. In addition, an undepleted pump wave is assumed with neglected
fiber losses and polarization influences. The determination of acoustic modes involved in a
obtained BGS is considered complex, due to the large number of possible intramodal and
intermodal interactions between all optical modes of the multimode fiber. Thus, for SBS
sensing in graded-index multimode fibers, the number of involved optical and acoustical
modes requires reduction to an absolute minimum.

However, there is no detailed analysis of intramodal and intermodal SBS in PFGI-POF as presented for silica fibers. Based on the comparable graded-index profile of silica multimode fibers and PFGI-POFs, intramodal and intermodal SBS are assumed to follow their silica equivalent. Furthermore, the Brillouin gain coefficient in PFGI-POF is expected to demonstrate a comparable dependency on the radial distance to the fiber core and angle of indecent, which is consistent with [45]. The velocity of the acoustic modes in PFGI-POF is expected to be significantly smaller compared to silica. Literature reports on acoustic velocities for bulk material (p-waves) in PMMA and polycarbonate to be around 2,750 m/s and 2,250 m/s, respectively, which are more than 2 times slower compared to silica ($v_a =$ 5,944 m/s). A significantly reduced velocity of acoustic modes leads to decreased spectral distances between adjacent acoustic resonances and therefore to less distortion of the overall BGS. On the other hand, the strong mode coupling of the PFGI-POF causes the excitation of neighbored higher-order optical modes. This return causes an additional distortion of the BGS. Thus, the key for SBS sensing in PFGI-POF is seen in the limitation of excited and evaluated optical modes.

CHAPTER **4**

Coupling between silica fibers and PFGI-POF

Designing optical communication links or FOS systems requires knowledge about the optical coupling efficiency. The upper limit of coupling efficiency is based on energy and radiance conservation principles [109]. Using the necessary boundary conditions on the source of radiation and the receiver to which it is coupled, the maximum achievable efficiency permits the evaluation of the optical system [109]. Since PFGI-POF are graded-index multimode fibers, only single mode and graded-index multimode fibers are discussed in this chapter. For communication links and Rayleigh backscattering FOSs a maximum power transmission through a coupling system is key. This is commonly realized by satisfying overfilled launch conditions [110]. In contrast, the generation of SBS requires an accurate coupling into the fundamental mode [18, 108, 111].

In general the coupling for an SBS sensor in graded-index MMFs needs to meet two requirements: (i) A maximum power transmission from the SMF of the setup into the fundamental mode of the fiber under test for SBS generation (see section 3.3). (ii) The maximum power launched from the fiber under test needs to be coupled on the photo detector to evaluate a BGS the most sensitive way. However, those two requirement are contradicting each other. In terms of (i) an underfilled launch condition and in terms of (ii) an overfilled launch condition needs to be satisfied. Thus, the coupling for SBS sensors in GI-MMFs is a compromise between both requirements. Further investigations are necessary to achieve the best possible coupling results.

Longitudinal, axial and angular optical misalignments lead to higher guiding modes within graded-index MMFs and therefore to an increased illuminated effective area resulting in a higher Brillouin threshold [111]. In addition, higher order modes are characterizes by greater propagation losses [112], which also raises the Brillouin threshold in silica fibers. However, the coupling efficiency into the fundamental mode of a graded-index MMF can not be determined by near and far field measurements [113]. Thus, a theoretical coupling description [114] forms the bases for an analytical model of coupling, describing the coupling coefficient.

The focus of this chapter is on the interconnection between PFGI-POFs and silica fibers. An analytical model of coupling is presented determining the insertion loss of an interconnection for each given optical mode. Lens-based and physical contact coupling of the PFGI-POF and its silica equivalent are discussed. The developed FC/APC connector used for all later investigations is specified.

4.1 Analytical description of coupling

Coupling coefficients

To describe any coupling effects, the definition of the Gaussian beam expressed in cylinder coordinates is needed:

$$E(r,z) \;=\; \frac{\omega(z)}{\omega_0} \cdot exp\left\{ \frac{-r^2}{\omega^2(z)} \right\} \cdot exp\left\{ \frac{jkr^2}{2R(z)+j\Phi(z)} \right\} \tag{4.1}$$

$$\omega(z) \;=\; \omega_0 \cdot \sqrt{1+\left(\frac{z}{z_0}\right)^2} \tag{4.2}$$

$$R(z) \;=\; z\left(1+\left(\frac{z}{z_0}\right)^2\right) \tag{4.3}$$

$$\Phi(z) \;=\; tan^{-1}\left(\frac{z}{z_0}\right) \tag{4.4}$$

$$z_0 \;=\; \frac{\pi\omega_0^2}{\lambda} \tag{4.5}$$

where ω denotes the beam diameter, ω_0 the diameter of the beam waist, r the radial distance from the center axis of the beam, k the wave number and z the axial distance from the beam's waist. $R(z)$ is the radius of curvature of the beam's wavefronts at the location z. Φ is the Gouy phase at the location z, which represents an extra phase term beyond that attributable to the phase velocity of light.

The transversal distribution of field amplitudes in a graded-index MMF is given by $TEM_{p,q}$ modes expressed in Eqn. 4.6. The transversal field distributions ψ_p and ψ_q are described by Hermite polynomials H and parameters of a Gaussian beam in Eqn. 4.7 and Eqn. 4.8 [109, 114–118]. Detailed descriptions of Hermite polynoms, Gaussian beams and the derivation of Hermite polynoms for mode field distributions are not given in this thesis.

$$TEM_{p,q} \;=\; \psi_p(x)\,\psi_q(y) \tag{4.6}$$

$$\psi_p(x) \;=\; \left(\sqrt{\frac{2}{\pi}}\frac{1}{\omega 2^p p!}\right)^{\frac{1}{2}} H_p\left(\frac{\sqrt{2}x}{\omega}\right) exp\left\{-\frac{x^2}{\omega^2} - jk\frac{x^2}{2R}\right\} \tag{4.7}$$

$$\psi_q(y) \;=\; \left(\sqrt{\frac{2}{\pi}}\frac{1}{\omega 2^q q!}\right)^{\frac{1}{2}} H_q\left(\frac{\sqrt{2}y}{\omega}\right) exp\left\{-\frac{y^2}{\omega^2} - jk\frac{y^2}{2R}\right\} \tag{4.8}$$

The field distribution of the fundamental mode in a graded-index MMF can be described as Gaussian shaped due to the parabolic graded-index profile of the fiber [115, 117]. In addition, the field distribution of a SMF is approximated by a Gaussian function [117]. Hence, both fiber modes are analytically described by Eqn. 4.6. Under the assumption that the numerical aperture of both fibers are wide enough to guide the incident light, a coupling coefficient between SMF and fundamental mode of the graded-index MMF can be calculated.

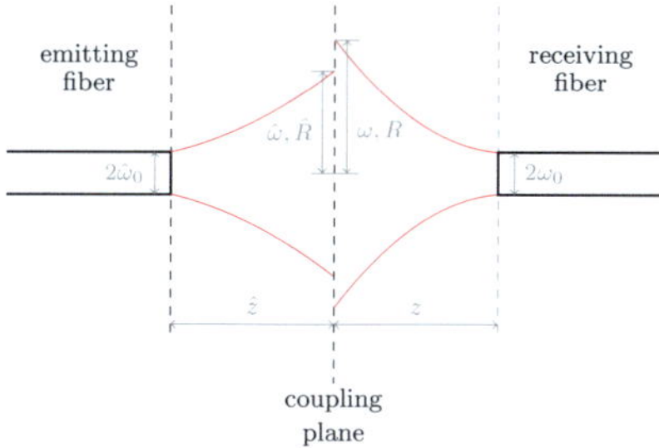

Figure 4.1: Schematic coupling between two single mode fibers. The Gaussian beam parameter are indicated in gray and the beam diameter propagating into the direction of the coupling plane in red.

In Fig. 4.1 the optical coupling between two fibers is displayed schematically. In the interest of legibility, the parameters of the light emitting fiber are marked as $\hat{p}$, $\hat{q}$, $\hat{\omega}$ and $\hat{R}$. At the fiber facets, the terms $exp\left\{-jk\frac{x^2}{2R}\right\}$ and $exp\left\{-jk\frac{y^2}{2R}\right\}$ of Eqn. 4.7 and Eqn. 4.8 can be neglected. The radiated field distribution of $\text{TEM}_{p,q}$ is projected on an imaginary plane between both fiber facets. This projection is transfered into the receiving fiber. The exact position of the plane can be chosen freely and there are no mode conversions assuming an accurate alignment. Hence, the radiated $\text{TEM}_{\hat{p},\hat{q}}$ modes stimulate optical modes in the second fiber described as $\text{TEM}_{p,q} = C_{\hat{p}\hat{q}pq} \cdot \psi_p(x) \cdot \psi_q(y)$. For the coupling plane between both fiber facets follows Eqn. 4.9 [117].

$$\psi_p(x)\psi_q(y) = \sum_p \sum_q C_{\hat{p}\hat{q}pq} \, \hat{\psi}_{\hat{p}}(x) \, \hat{\psi}_{\hat{q}}(y) \tag{4.9}$$

Benefiting from the orthogonality of both field distributions of each fiber facet, the coupling coefficients $C_{\hat{p}\hat{q}pq} = c_{\hat{p}p}c_{\hat{q}q}$ are determined by integrating them over the surface of the coupling plane:

$$c_{\hat{p}p} = \int_{-\infty}^{\infty} \hat{\psi}_{\hat{p}}(x)\psi_p^*(x) \, dx \tag{4.10}$$

$$c_{\hat{q}q} = \int_{-\infty}^{\infty} \hat{\psi}_{\hat{q}}(y)\psi_q^*(y) \, dy \tag{4.11}$$

When inserting the definition of ψ_p Eqn. 4.7 into Eqn. 4.10 follows [117]:

$$c_{\hat{p}p} = \left(\frac{2}{\pi \hat{\omega} \omega \hat{p}! p! \cdot 2^{\hat{p}+p}} \right)^{\frac{1}{2}} \int_{-\infty}^{\infty} H_{\hat{p}} \left(\frac{x\sqrt{2}}{\hat{\omega}} \right) H_p \left(\frac{x\sqrt{2}}{\omega} \right) exp\left\{ -\Psi x^2 \right\} dx \quad (4.12)$$

$$\text{with} \quad \Psi = \frac{1}{\hat{\omega}^2} + \frac{1}{\omega^2} + j\frac{k}{2} \left(\frac{1}{\hat{R}^2} - \frac{1}{R^2} \right)$$

Equation 4.12 gives the condition, that does not allow for there to be any coupling between odd and even mode numbers. Furthermore, solving Eqn. 4.12 results in the following coupling coefficients [117]. Please note, that the following equations are a selection of solutions for chosen values of p and q:

$$m = \frac{1}{\hat{\omega}_0^2 - j\frac{\lambda \hat{z}}{\pi}} + \frac{1}{\omega_0^2 - j\frac{\lambda z}{\pi}} \quad (4.13)$$

$$c_{00} = \sqrt{\frac{2}{\hat{\omega}\omega m}} \quad (4.14)$$

$$c_{02} = \frac{c_{00}}{\sqrt{2}} \left(\frac{\hat{\omega}}{\omega} c_{00}^2 - 1 \right) \quad (4.15)$$

$$c_{20} = \frac{c_{00}}{\sqrt{2}} \left(\frac{\omega}{\hat{\omega}} c_{00}^2 - 1 \right) \quad (4.16)$$

$$c_{22} = \frac{1}{c_{00}} \left(c_{00}^6 + c_{02}c_{20} \right) \quad (4.17)$$

$$c_{11} = c_{00}^3 \quad (4.18)$$

$$c_{13} = \sqrt{3} \, c_{00}^2 c_{02} \quad (4.19)$$

$$c_{31} = \sqrt{3} \, c_{00}^2 c_{20} \quad (4.20)$$

$$c_{33} = c_{00} \left(c_{00}^6 + 3c_{02}c_{20} \right) \quad (4.21)$$

Fiber parameter

Table 4.1: Fiber parameters used for calculations. Note that the mode field radii of the multimode fibers were calculated for fundamental mode only. [119]

Fiber type	Fiber name	Core radius a [μm]	Numerical aperture	Fiber parameter V	Mode radius ω_0 [μm]
SMF	SMF-28	4.1 [120]	0.14 [120]	2.734	4.16
GI-MMF	ClearCurve	25 [83]	0.20 [83]	23.82	7.29
PFGI-POF	GigaPOF-50SR	25 [82]	0.19 [82]	22.63	7.49

For all the following calculations the fiber parameters of Tab. 4.1 were applied. All parameters are related to a wavelength of $\lambda = 1319$ nm. The fiber parameter V was calculated by using the relative refractive index difference Δn between core and coating as well as the core diameter a of the fiber (see Eqn. 4.22). The mode field diameter ω_0 was computed by using V and a (see Eqn. 4.23 [121] and Eqn. 4.24 [122] for graded-index and single mode fibers, respectively).

$$V = \frac{2\pi a}{\lambda} \Delta n = \frac{2\pi}{\lambda} a \sqrt{n_2^2 - n_1^2} \quad (4.22)$$

$$\frac{\omega_0}{a} = 0.65 + \frac{1.619}{V^{3/2}} + \frac{2.879}{V^6} \quad (4.23)$$

$$\frac{\omega_0}{a} = \sqrt{\frac{2}{V}} + \frac{0.23}{V^{3/2}} + \frac{18.01}{V^6} \quad (4.24)$$

Applying the fiber parameter

Based on the coupling coefficients and the mode field radii from Tab. 4.1 mode conversions between the fibers can be computed. The power coupled from $\text{TEM}_{\hat{p},\hat{q}}$ to $\text{TEM}_{p,q}$ is expressed by $|c_{\hat{p},p}|^2 \cdot |c_{\hat{q},q}|^2$. Furthermore, the mode conversions caused by misalignments can be calculated. If there is no misalignment the TEM modes of the emitting fibers are equal to the TEM modes of the receiving fiber. In that case all coupling coefficients are zero except for $c_{ii} = 1$ with $i = 0, 1, 2, 3, \dots$.

Longitudinal misalignment

As shown in Fig. 4.1 the distance between both fiber facets is expressed by $\hat{z} + z$. Since the Gaussian beam diameter ω is only a function of distance (see Eqn. 4.2), the coupling coefficients can be deployed. In case of coupling from a fundamental mode to another fundamental mode the transmitted power is given by Eqn. 4.25 [118, 119] and plotted in Fig. 4.2. The power coupling coefficient κ is proportional to a mode-based insertion loss next to neglected Fresnel reflections [114].

$$\kappa = c_{00} \cdot c_{00}^* = \frac{4}{|\hat{\omega}\omega m|^2} = \frac{4}{\left(\frac{\omega_0}{\hat{\omega}_0} + \frac{\hat{\omega}_0}{\omega_0}\right)^2 + \left(\frac{\lambda}{\pi\omega_0\hat{\omega}_0}\right)^2 + (\hat{z} + z)^2} \tag{4.25}$$

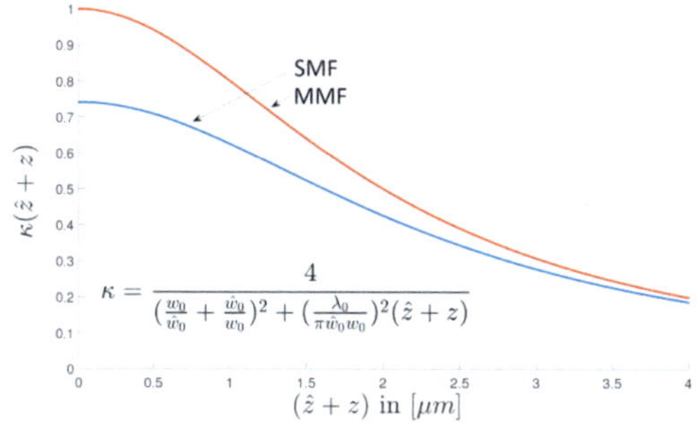

Figure 4.2: GI-MMF fundamental mode to fundamental mode coupling power coefficients κ for longitudinal misalignments. The plots "SMF" and "MMF" are representing the receiving fiber types. To improve clarity, the fibers PFGI-POF and GI-MMF are combined as "MMF" due to the graphs almost progressing the same way. [119]

Parts of the power of the emitted fundamental mode stimulate higher order modes in the receiving fiber. This mode conversion is represented by 2 chosen values for $\hat{z} + z$ in Fig. 4.3. It is note, that only even mode numbers are excited, which is due to symmetry in the Hermite polynomials. Furthermore, the number of stimulated modes is observed to increase, whereas the value of each coupling coefficient decreases with growing longitudinal misalignments.

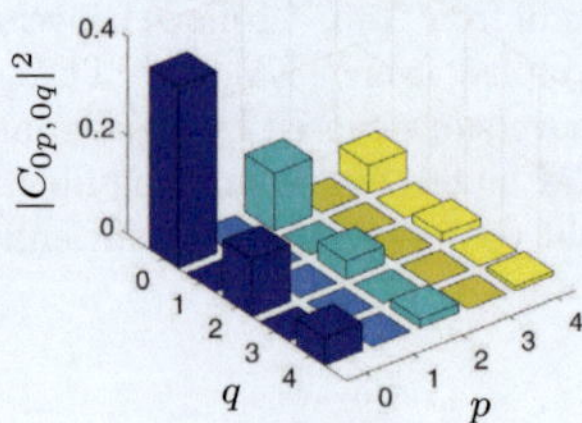
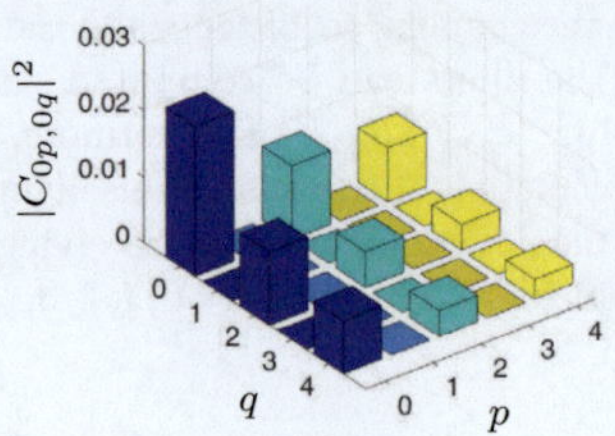

Figure 4.3: Coupling coefficients for higher order modes by applying a longitudinal mis-alignment of 365 μm (50 times ω_0, left) and 1822 μm (250 times ω_0, right) for GI-MMF to GI-MMF [118, 119].

Axial misalignment

Since fibers are rotationally symmetric only axial misalignments in the direction of x are described. All parameters regarding the direction of y remain the same. In addition, all axial misalignments are described for the emitting fiber. A schematic axial misalignment is shown in Fig. 4.4. The field distribution of the fundamental mode $\psi_0(x)$ changes to Eqn. 4.26 [118].

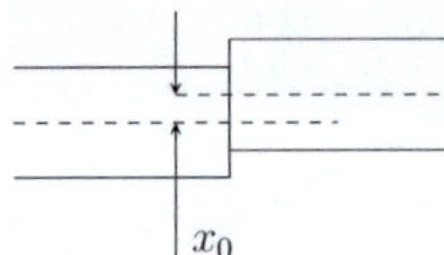

Figure 4.4: Schematic coupling between two fibers with axial misalignment. x_0 represents the offset between the optical axes of both fibers. [118]

$$\hat{\psi}_0(x) = \left(\sqrt{\frac{2}{\pi}}\frac{1}{\hat{\omega}}\right)^{\frac{1}{2}} exp\left\{-\frac{(x-x_0)^2}{\hat{\omega}^2} - jk\frac{(x-x_0)^2}{2\hat{R}}\right\} \tag{4.26}$$

Hence, the field and power coupling coefficients are changing to Eqn. 4.27 and Eqn. 4.28, respectively. It should be taken into account, that only the first term of the power series is given for κ_{x0}. As for longitudinal misalignments, κ_{x0} is plotted as a function of axial offset in Fig. 4.5.

$$c_{x0} = c_{00}\, exp\left\{-\frac{x_0^2}{m}\left(\frac{1}{\hat{\omega}^2}+\frac{jk}{2\hat{R}^2}\right)\left(\frac{1}{\omega^2}-\frac{jk}{2R}\right)\right\} \tag{4.27}$$

$$\kappa_{x0} = |c_{x0}|^2|c_{00}|^2 = \kappa - \frac{1}{2}x_0^2\left(\frac{1}{\hat{\omega}_0^2}+\frac{1}{\omega_o^2}\right) \tag{4.28}$$

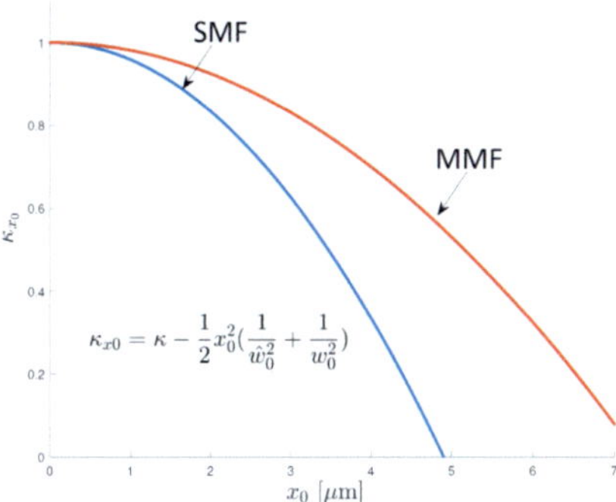

Figure 4.5: GI-MMF fundamental mode to fundamental mode coupling power coefficients κ_{x0} for axial misalignments. The plots "SMF" and "MMF" are representing the receiving fiber types. To improve clarity, the fibers PFGI-POF and GI-MMF are combined as "MMF" as the graphs almost progress the same. [119]

Axial misalignments for higher order modes can be estimated by substituting c_{00} with c_{x0} for the emitting fiber. Figure 4.6 shows a selection of modes coupled between to GI-MMFs for given axial misalignments. Only one index is given for each fiber, due to the fact that the fiber core is rotationally symmetric. In contrast to the longitudinal misalignment, a mode mixing within even and odd mode numbers can be observed. The bigger the axial misalignment gets, the more modes in the receiving fiber are excited and the less power each mode carries.

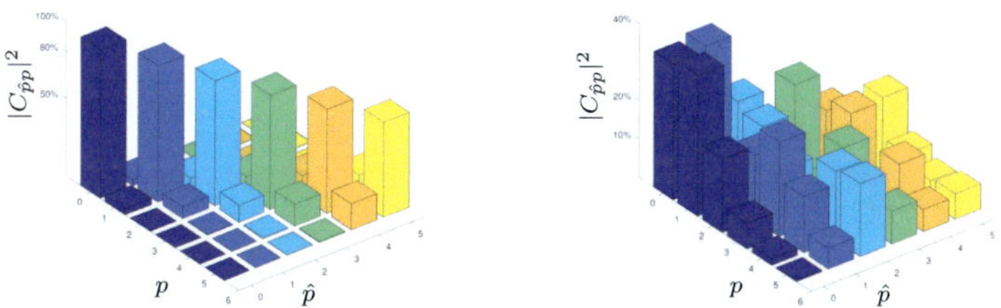

Figure 4.6: Coupling coefficients for higher order modes by applying an axial misalignment of 1μm (left) and $5\ \mu$m (right) for GI-MMF to GI-MMF [118, 119].

Angular misalignment

The angular misalignment is schematically shown in Fig. 4.7. As for axial misalignments, the fiber parameters are described in the direction of x for the emitting fiber. Using the small-angle approximation $sin\theta \approx \theta$ the fundamental mode field distribution of the emitting fiber is given in Eqn. 4.29. $k\theta x$ represents the projection of the wave vector $\mathbf{k}$

on the coupling plane. Consequently, the field and power coupling coefficients change to Eqn. 4.30 and Eqn. 4.31, respectively.

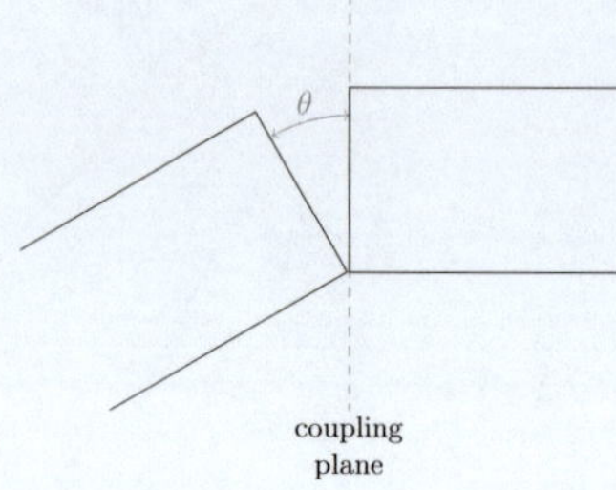

Figure 4.7: Schematic coupling between two fibers with angular misalignment. θ represents the angle between both fiber facets.

$$\hat{\psi}_0(x) = \left(\sqrt{\frac{2}{\pi}}\frac{1}{\hat{\omega}}\right)^{\frac{1}{2}} exp\left\{-\frac{x^2}{\hat{\omega}^2} - jk\frac{x^2}{2\hat{R}} - jk\theta x\right\} \tag{4.29}$$

$$c_\theta = c_{00}\, exp\left\{\frac{-\kappa^2\theta^2}{4m}\right\} \tag{4.30}$$

$$\kappa_\theta = |c_\theta|^2|c_{00}|^2 = \kappa - \frac{1}{2}\kappa^2\theta^2(\hat{\omega}^2 + \omega^2) \tag{4.31}$$

The coupling efficiency between both fundamental modes κ_θ is plotted as a function of angular misalignment in Fig. 4.8. Figure 4.8 demonstrates a strong declining coupling efficiency for very small angular misalignments.

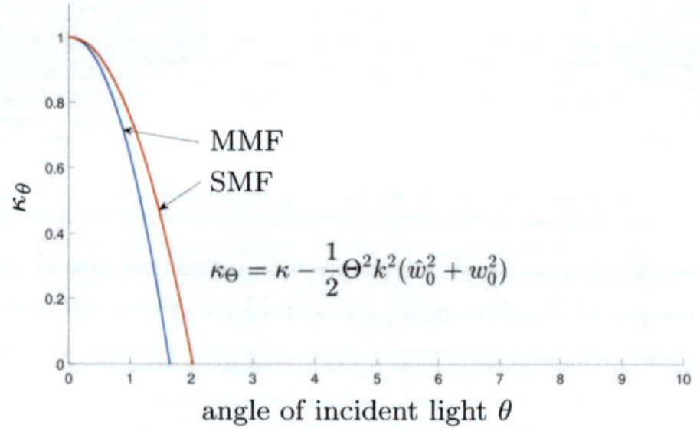

Figure 4.8: GI-MMF fundamental mode to fundamental mode coupling power coefficients κ_θ for angular misalignments. The plots "SMF" and "MMF" are representing the receiving fiber types. To enhance clarity, the fibers PFGI-POF and GI-MMF are combined as "MMF" as the graphs almost progress the same. [118].

Discussion of results

Fundamental mode coupling efficiencies were demonstrated for given types of misalignments. A decrease of fundamental mode coupling efficiency was calculated for increasing fiber misalignments. In addition, the number of stimulated modes in a receiving GI-MMF was shown to increase for greater values of misalignments which can be described as mode conversation. In general the coupling between two GI-MMFs can be considered less affected by misalignments compared to SMF-SMF or even SMF-GI-MMF, due to bigger mode field radii. Angular misalignments are seen critical for fiber interconnections, whereas longitudinal misalignments are having a modest impact. However, a combination of all misalignment types results in greater decreased coupling efficiencies as the sum each type calculated separately. The acquired knowledge is put into practice by defining an 1 dB loss limit for each type of misalignment that leads to the following boundary conditions for SMF to PFGI-POF (a $= 50$ μm) coupling: $\hat{z} + z = 1.2$ μm, $x_0 = 2.1$ μm and $\theta = 1.1$ degrees. These imposed limitations ensure the predominant illumination of the PFGI-POF facet corresponding to a fundamental mode propagation along the fiber.

4.2 Coupling based on lenses

Fiber coupling based on lenses has been studied and applied for several decades [123–125]. The most commonly known lens-based fiber coupling into POF named TOSLINK was developed by Toshiba in 1983 and became an addition to IEC958 1989-03 in 1989 [112]. In terms of PFGI-POFs ball-lensed fiber interconnections were presented [110, 126] and will be commercially available in the nearer future.

The main idea of lens-based coupling is the adjustment of mode field diameters for different fiber types. A $4f$-setup with ball-lenses is schematically displayed in Fig. 4.9. Every optical lens system used for fiber interconnection can be simplified as a coupling plane, irrespective of number and types of lenses. Hence, the coupling coefficients for fiber modes also apply for lens-based coupling approaches.

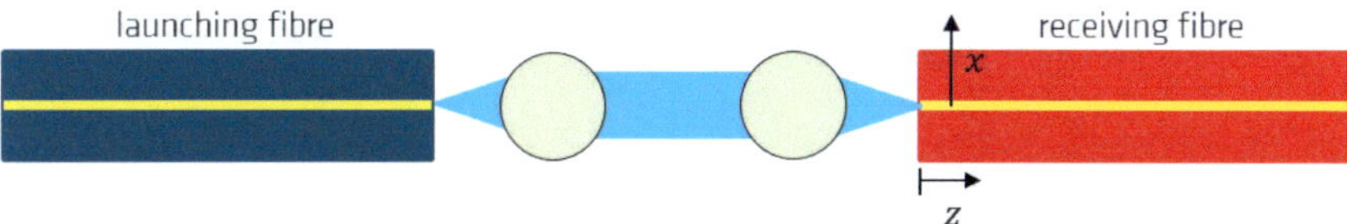

Figure 4.9: Schematic view on a ball-lensed coupler. Longitudinal and axial misalignments are represented by z and x, respectively [119].

Next to the coupling coefficients Fresnel reflections must be taken into consideration for lens-based coupling. As shown in Fig. 4.9, there are four air-lens interfaces and two fiber-air interface causing reflections. Assuming the Gaussian beam diameter is significantly smaller than the ball-lense diameter of the surface, the lens can be seen as a plain surface perpendicular to the optical axis of the fiber. In this case, Fresnel reflections are calculated irrespective of the angle of light incidence as given in Eqn. 4.32 [3]. n_a and n_b are the refractive index of the two interacting media at the interface.

$$R_p = \left| \frac{n_a - n_b}{n_a + n_b} \right|^2 \tag{4.32}$$

The refractive indexes for the most common lens materials were calculated for a wavelength of 1319 nm using the Sellmeier equation. The Sellmeier equation is given in Eqn. 4.33 together with its coefficients in Tab. 4.2. Based on the determined refractive indexes the minimum insertion of $4f$ ball-lens system for given lens material is given in Tab. 4.3.

$$n^2(\lambda) = 1 + \frac{B_1 \lambda^2}{\lambda^2 - C_1} + \frac{B_2 \lambda^2}{\lambda^2 - C_2} + \frac{B_3 \lambda^2}{\lambda^2 - C_3} \tag{4.33}$$

Table 4.2: Table of coefficients for Sellmeier equation used to calculate the refractive index. [127]

Material	B_1	B_2	B_3	C_1 $[\cdot 10^{-3}\ \mu m^2]$	C_2 $[\cdot 10^{-2}\ \mu m^2]$	C_3 $[\mu m^2]$
borosilicate crown glass (BK7)	1.0396	0.2318	1.0105	6.00070	2.00179	103.561
sapphire (ordinary wave)	1.4314	0.6506	5.3414	5.27993	1.42383	325.018
sapphire (extra-ordinary wave)	1.5040	0.5507	6.5927	5.48041	1.47994	402.895
fused silica	0.6962	0.4079	0.8975	4.67915	1.35121	97.9340

Table 4.3: Minimum insertion loss in dB for various lens materials and connector configurations including two SOF-air interfaces. The refractive index for each given lens material was determined by the Sellmeier equation for a wavelength of 1319 nm.

lens material	1 lens	2 lenses
Fused silica	2.000	2.485
NBK-7	2.099	2.710
Sapphire	2.606	3.979

For the purpose of calculating a mode dependent insertion loss, the values of Tab. 4.3 have to be multiplied with the mode dependent coupling coefficients corresponding to the optical modes of interest. This procedure allows for the calculation of all given coupling constellations of lenses and fibers as well as for given misalignments.

A detailed description of ball-lensed coupling between PFGI-POF and silica optical fibers is given in the master thesis of Tilman Jeschke [118] supervised by the author.

4.3 Physical contact coupling

Almost all fiber optic components are connected via physical contact connectors of numerous specifications (LC, ST, FC, E2000, SMA, ...). All in common is the prevalent factor of polished fiber facets are touching each other only at their cores, allowing transmissions with low loss. The fiber end is embedded in a ferrule and spring-loaded to control the force as the plug is screwed into the receptacle. Hence, longitudinal and angular misalignments are suppressed. However, standardized or commercially available physical contact connectors for PFGI-POFs are non-existing. All PFGI-POF connectors have been hand-crafted by each research group individually and are based on physical contact [18, 21, 45, 94, 128], apart of the ball-lensed connectors shown in [110].

Furthermore, the coupling coefficients are valid for physical contact connectors as the coupling plane can be applied at the fiber facet of the receiving fiber. Fresnel reflections have to be taken into account for physical contact connectors, too. Based on Eqn. 4.32, the reflection-induced transmission loss with air between the fiber interfaces are 0.25 dB and 0.30 dB for SOF-POF and SOF-SOF, respectively.

As for the lens-based coupling there have been approaches adjusting the mode field diameters of SMF and PFGI-POF by inserting multimode fibers [129]. To ensure an accurate fiber core alignment between single mode fibers and PFGI-POF, the monitoring of the BGS was proposed [111].

Since fiber facets are touching each other to ensure low transmission losses, the roughness of the PFGI-POF facet is of special interest. PFGI-POFs are characterized by lower stiffness and lower elastic modulus compared to PMMA-based POFs and silica optical fibers. Thus, the fiber facet preparation is the essential process step in connector manufacturing. Figure 4.10 shows a PFGI-POF glued into a metallic substrate and manually polished to the same level of height as the metal substrate using 0.3 μm grit calcined alumina polishing sheets. However, this PFGI-POF facet did not support light guidance into the fiber.

Using the Keyence VHX-6000 digital microscope the fiber interface roughness was studied more closely. The PFGI-POF was observed to form a concave depression (approximately 16 μm at its lowest point) with local grooves (up to 26 μm) (see upper right and lower image in Fig. 4.10). Metal particles removed by the polishing grit were not detected on the fiber facet. The concave formed depression can be associated to the differences in hardness of PFGI-POF and metal substrate. The polishing grit is predominantly seen to remove parts of the fiber at first, followed by a smoothening of the metal substrate surface. The local grooves can be related to a piecemeal detachment of fiber parts. The concave depression and the grooves can be considered to act as centers of scattering and reflections for the incident light. This however, is dependent on the structure caused by the polishing grit. Consequently, a reproducible fiber connector based on physical contact can not be crafted with the guarantee of a stable coupling into the fundamental mode.

A simple fiber interconnection with the the aim of a short production time and a high demand for manageability lead to the physical contact connector approach.

Non-angled FC connectors employed in [18] demonstrated to have a strong dependency on temperature on reflected power. Thus, FC/APC connector were chosen to reduce the power of Fresnel reflections by an 8 degree angled cut of the fibers. A FC/APC fiber connector blank was externally manufactured from stainless steel with a 510 + 5/ -0 μm centered hole. The deviation of the center position was measured to be $\pm$ 2 μm. The data sheet of the PFGI-POF provides the following geometrical uncertainties: Core concentricity ($\leq$ 4 μm [82]), core diameter (50 $\pm$ 5 μm [82]), fiber outer diameter (490 $\pm$ 5 μm [82]) and

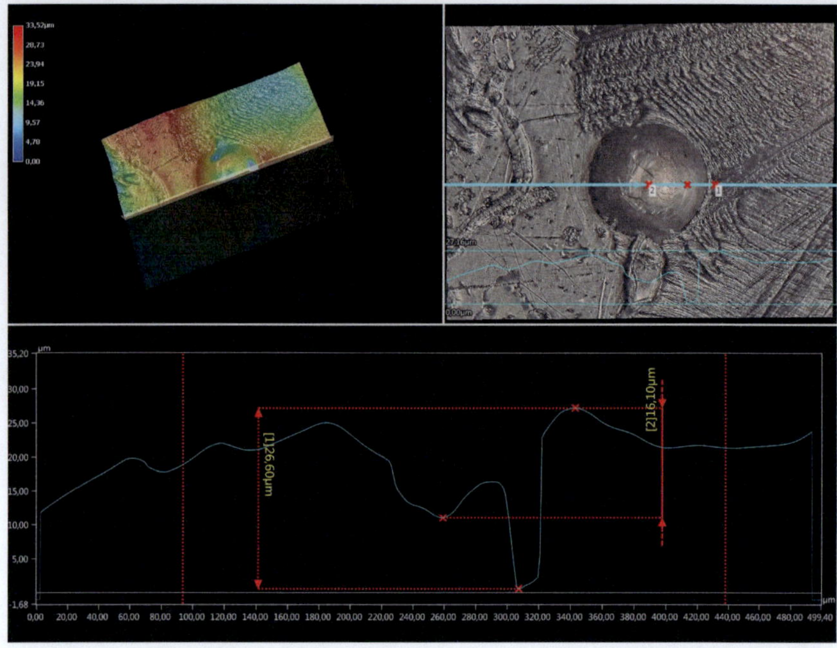

Figure 4.10: Microscopic image and evaluation of a polished PFGI-POF facet. Upper left: 3D landscape of connector image. Upper right: Microscopic image of fiber coating and core compared to a metallic substrate. Lower: Height measurement of surface roughness along the line drawn in the upper right image.

numerical aperture (0.19 ± 0.015 [82]). However, this deviation of the core center position is unknown. Due to all of the named manufacturing uncertainties, the axial misalignment error can be considered as the dominating effect reducing the fundamental mode coupling efficiency as the fiber connector minimizes angular and longitudinal misalignments by design.

The issue of the structured PFGI-POF facet can be solved by leaving a fiber supernatant on the connector surface (approximately 100 μm) after polishing. Since the PFGI-POF has a lower elastic modulus compared to silica, stainless steel and zirconia ceramic (ferrule material of silica optical fibers [130]), the supernatant of the PFGI-POF is squeezed on to the fiber facet of the silica fiber. This compression compensates for the surface roughness and avoids on additional scattering effects leading to increased transmission losses. A compression-induced change of PFGI-POF parameters can not be excluded. Those changes are assumed to be negligibly small, as SMF-PFGI-POF interconnections employed on bases of 3-axis stages demonstrated to have no significant differences in insertion loss and SBS backscattering power. The compression force load is provided by the constant force springs employed in all FC/APC connectors, while the PFGI-POF is fixed with glue into the stainless steel connector blank.

The insertion losses measured at the SMF-PFGI–POF and PFGI-POF–SMF interfaces of the physical contact connector were 0.14 dB and 7.67 dB, respectively [21]. The significant low insertion loss from SMF to PFGI-POF predominantly originates from Fresnel reflection due to the refractive indices. In the opposite direction the insertion loss should be considered as a superposition of modal selection, mismatched mode field diameters between the fibers and Fresnel reflection. On the other hand, the modal selection ensures the evaluation of backscattered light that predominantly originates from the fundamental mode of the PFGI-POF.

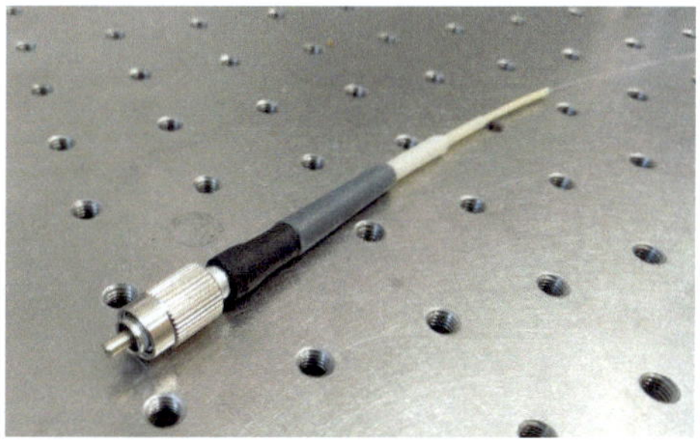

Figure 4.11: Hand-crafted FC/APC connector for PFGI-POFs.

Using FC/APC mating sleeves, each PFGI-POF was permanently connected to a SMF patch cord. This interconnection provided easy handling and connectivity with laboratory setups and devices. All used SMF-PFGI-POF connections in this work were crafted with a self developed kink protection. The FC/APC connector for PFGI-POFs is displayed in Fig. 4.11.

CHAPTER 5

Distributed Brillouin sensors

In 1976 M.K. Barnoski and S.M. Jensen presented the basic concept of the nowadays known optical time-domain reflectometry (OTDR) [131]. Pulsed light from a laser was coupled into a glass fiber and the time dependence of the light backscattered within the fiber as the pulse travels down the waveguide was recorded. While traveling along the fiber, the light pulse encounters spots of scattering and reflection (faults or cracks in the fiber or its end face). Consequently, portions of the injected light will be backscattered from these spots. In addition, portions of the pulse power are continuously sent back to its point of origin while traveling along the fiber due to the Rayleigh scattering. The intensity of the received light is recorded according to its time of arrival, which provides the temporal impulse response of the fiber $g(t)$. Based upon $g(t)$ a spatially distributed profile of reflection intensities $g(z)$ can be computed following the Eqn. 5.1. At $t = 0$, the light pulse is launched into the fiber and travels with the velocity c/n.

$$g(z) = \frac{1}{2}\frac{c}{n}g(t) \tag{5.1}$$

Localizations along a fiber can be further achieved through scanning of the probing optical frequency (OFDR - optical frequency-domain reflectometry) or through correlation techniques utilizing frequency modulated light waves (OCDR - optical correlation-domain reflectometry). The adoption of all named localization methods for non-linear backscattering effects, such as stimulated Brillouin scattering, has been an intense process. In this chapter, a basic overview is given on distributed SBS sensing separated according to the domain approach. The Brillouin optical frequency-domain analysis and its theoretical background are presented in detail. Finally, a literature survey on SBS sensors in PFGI-POF is provided.

5.1 Brillouin optical time-domain approach

Translating the basic principal of an OTDR on stimulated Brillouin backscattering resulted in the 1989 introduced Brillouin optical time-domain analysis (BOTDA) [132]. According to chapter 2 and [132], BOTDA is based on the interaction between a pulsed pump wave and a counter-propagating continuous wave probe. Both light paths can be tuned in frequency to each other. Due to the electrostriction of the optical beating caused by the two optical waves, an acoustic wave is locally induced at the point where the pump pulse and the continuous probe meet. Depending on the frequency detuning of the optical waves to each other and the phonon frequency of the fiber under test, the probe wave amplification varies. The maximum amplification is observed when $f_p - f_s = f_a$, where f_p, f_s and f_a are the pump, probe, and phonon frequency, respectively. Thus, a stepwise frequency detuning around the phonon frequency enables the spectral measurement of spatial fiber segments. The determination of BFS at each fiber segment is the result of a fitting process based on a theoretical profile describing the measured BGS.

The fiber segment illuminated by the pump light pulse and therefore the Brillouin interaction length, can be estimated utilizing Eqn. 5.1. Taking into account the pump pulse duration t_d, the illuminated fiber length Δz follows $\Delta z = c\, t_d/2n$. Thus, Δz can be seen as the spatial resolution of a conventional BOTDA system. In principle, the shorter the pump pulses get, the more spatial segments can be theoretically evaluated. However, the spatial resolution is limited to only 1 m (corresponding to a pulse width of 10 ns) in silica optical fibers by two predominant factors. First, reduced pulse durations are shorten the Brillouin interaction length. As presented in chapter 2, the SBS gain scales with the Brillouin interaction length, which leads to weaker backscattering intensities for shorter pump pulses. Consequently, the estimation of the BFS is more sensitive to noise. Secondly, shorter pump pulses have broader power spectra. Thus, the effective Brillouin gain coefficient needs to be described as the convolution of pump spectrum density and the natural Brillouin linewidth [133]. For pump pulses shorter than 20 ns, the BGS linewidth is approximately given by $1/t_d$ [11]. The inherent broadening of the BGS causes a reduced gain of all detuning frequencies, which in addition reduce the signal-to-noise ratio.

Based on the presented basic BOTDA system numerous issues have been solved such as polarization fading [134], laser incoherence [135, 136], modulation instability [137], nonlocal effects in probe wave modulation [138] and pump depletion [139]. Further enhancements of the setup have been presented with the aim to cover longer sensing ranges by using discrete amplification [140], pump coding [141], distributed Raman amplification [14, 142] and self-heterodyne detection [143]. Furthermore, higher spatial resolutions down to 2 cm [144, 145] have been achieved by employing methods such as differential pulse-width pairs (DPP-BOTDA) [146], pump pre-pulses (PPP-BOTDA) [147], gain-profile tracing (GPT-BOTDA) [145], bright pulses [148], dark pulses [144] and Walsh coded bright pulses [149]. Improvements regarding the measuring time have been achieved by introducing schemes like slope-assisted BOTDA [150] and sweep-free multi-tones [151]. Next to BOTDA which requires access to both ends of the sensing fiber, a single end access derivative has been presented and named Brillouin optical time-domain reflectometry (BOTDR) [64]. Employing the parasitic light originating from Rayleigh backscattering and Fresnel reflections from connectors, spontaneous Brillouin backscattering can be evaluated. Reference [152] summarizes the current state of the art techniques employed in BOTDR addressing the same issues as for BOTDA. To resolve small scale spatial events the spatial resolution is enhanced by double-pulse BOTDR (DP-BOTDR) [153], a synthetic BOTDR

(S-BOTDR) [154], a phase shift pulse BOTDR (PSP-BOTDR) [155] and a differential cross spectrum BOTDR (DCS-BOTDR) [156]. In order to enhance the measurement speed in BOTDA and BOTDR slope-assisted approaches have been presented [157, 158] . However, the reduction of measurement time develops another limitation concerning the dynamic range. To avoid this limitation, multiple slopes have were evaluated in [159, 160]

5.2 Brillouin optical correlation-domain approach

Compared to BOTDA and BOTDR, the Brillouin optical correlation-domain analysis (BOCDA) achieved a more precise higher spatial resolutions of a few millimeter [161]. BOCDA was first presented in 2000 [162]. In principle, BOCDA is based on the interaction of two sinusoidally frequency-modulated and counter-propagating light paths. For constant frequency differences between the pump and probe wave, selected positions along the fiber can face the matching phonon frequency of the fiber under test. This results in a SBS amplified probe wave along a spatially short distance. The correlation peaks of the received increased backscattering signals are called nodes. When choosing the appropriate frequency f_m and frequency amplitude f_{hub} of the modulations with respect to the phonon frequency of the fiber under test, a localization along the fiber is achieved. Notably, a complete and stable buildup of the acoustic field is assumed for the SBS interaction [11, 162]. Further to that, the width of the measured BGS should ideally be as narrow as it can be obtained from a constant wave interrogation [11]. However, BOCDA poses increased demands on the employable frequency range of the modulators and the phase noise of utilized frequency generators [163]. Furthermore, the point-by-point spatial scanning, on top of the required frequency detuning around the phonon frequency of the fiber is quite time consuming [11]. In return and in contrast to BOTDA, the spatial resolution Δz of this method is expected to be related to the BGS linewidth Δf_B since there is no appreciable buildup of the acoustic field once the pump-probe frequency difference exceeds [11]. The spatial resolution is given by [162]:

$$\Delta z = \frac{c \cdot \Delta f_B}{n \cdot 2\pi \cdot f_m \cdot f_{hub}}.$$

Generally, the avoidance of simultaneous Brillouin interactions from multiple points along the fiber is key in BOCDA. Therefore, the fiber length L must remain shorter than the distance of neighbored obeying nodes obtained by the correlation. Reference [164] gives the maximum fiber length L_{max} in BOCDA systems as $L_{max} \equiv c/(n \cdot 2f_m)$. Thus, there is a trade-off between sensing fiber length and spatial resolution.

Further improvements have been presented to overcome limiting issues. To remove co-existing correlation peaks additional time-domain gatings were introduced demonstrating a 1 km sensing range with 7 cm spatial resolution [165]. In order to increase measurement speed enhancements have been proposed based on slightly different frequencies in the pump and probe path [166] or by utilizing a voltage-controlled oscillator for faster probe wave sweeping and employing arbitrary multiple sensing points along a fiber [167]. Aiming to reduce the disturbance from points away from the correlation peaks, post-processing methods have been demonstrated to show significant enhancements [168].

Within the last few years phase-modulated BOCDA has been intensively studied. In principle, a fast randomly changing phase modulation is utilized rather than sinusoidal frequency modulation for the localization along the fiber [169]. This BOCDA derivate employs pseudo-random bit sequence modulated on the phase of pump and probe path. Thus, as long as these two counter-propagating phases overlap, they contribute to the creation of a stable acoustic field at a specific section in the fiber under test. Due to the characteristic of the pseudo-random code the Brillouin interaction occurs only along one segment of fiber. The sensing segment can be spatially moved along the fiber by controlling the delay between the instances the random sequences are launched into the pump/probe ends of the fiber [11]. For bit durations of 100 ps, a spatial resolution of 1.7 cm was demonstrated [169]. Though, the spatial region of sustained Brillouin amplification is governed by the spatial extent of the autocorrelation function of the sequence [169]. As a consequence, non-zero side-lobes of the auto-correlation function contaminate the probe with unwanted gain contributions. This issue has been solved by employing Golomb codes [170]. Hybrids of BOTDA and phase-modulated BOCDA have demonstrated impressive results such as a 17.5 km sensing range with 14 mm resolution [171]. However, the required measurement time is definitely a downside. BOCDA setups focusing on extreme small-scale spatial resolution down to the micro-scale are summarized in [17]. Also slope-assisted [157] and double slope-assisted [172] derivations have been proposed to achieve higher measurement speed. However, the slope-assisted approach drastically limits the detectable dynamic range of effects such as strain and temperature.

As for BOTDA, single ended variants of BOCDA have been proposed such as linear-configured BOCDA (L-BOCDA) [173] and Brillouin optical correlation-domain reflectometry (BOCDR) [48]. L-BOCDA is characterized by injecting pump and probe path at the same end of the under test. The counter-propagation of pump and probe path are predominately achieved by Fresnel reflection of the fiber end. A sensing range of 40 m with a spatial resolution of 16 cm has been demonstrated [173]. In contrast to L-BOCDA, BOCDR is based on spontaneous Brillouin scattering [48,50]. Utilizing a temporal gating scheme, 66 cm spatial resolution and 1 km measurement range were obtained with 50 Hz sampling rate [174]. By reducing the fiber length higher sampling rates up to 100 Hz and smaller-scaled spatial resolutions of 39 cm have been demonstrated [49]. In the last recent years the sloped-assisted BOCDR has came to the fore to provide a dynamic measurement method with high spatial resolution combined with intrinsic one-end accessibility. Slope-assisted BOCDR was first introduced in 2016 [175]. The power of incident light was shown to affect the achieved spatial resolution [176]. The sensitivity was shown to be enhanced with higher incident power and/or lower spatial resolution. Furthermore, a trade-off between the strain dynamic range and spatial resolution needs to be considered [177] as for slope-assisted BOCDA techniques. By employing a polarization maintaining SMF as fiber under test, high stability and a 1.4 times increased sensitivity compared to standard SMFs has been achieved within a sensing range of 10 km [178].

5.3 Brillouin optical frequency-domain approach

The Brillouin optical frequency-domain analysis (BOFDA) was introduced in 1996 [179]. BOFDA follows the frequency scanning scheme of counter-propagating pump and probe waves as described for BOTDA and BOCDA. The frequency difference between pump and probe wave is symboled f_D. The most evident difference lies in the concept of localization along the fiber. BOFDA takes advantage of the reversibility between the time and frequency domain given by a Fourier transform in the analysis of linear systems [180].

By measuring the complex transfer function $G(jf)$ with a vector network analyzer (VNA), the impulse response $g(t)$ of the fiber under test obeys the inverse fast Fourier transform (iFFT) of $G(jf)$. Consequently, following Eqn. 5.1 enables the localization along the fiber under test. The whole process is given in Eqn. 5.2.

$$G(jf, f_D) \bullet\!\!-\!\!\circ g(t, f_D) \xrightarrow[z=\frac{1}{2}\frac{c}{n}t]{} g(z, f_D) \tag{5.2}$$

BOFDA does not employ optical pulses but their spectral equivalents. In contrast to BOTDA, iFFT transformed pulse widths are arbitrary corresponding to arbitrarily spatial resolutions. The narrow bandwidth filtering of BOFDA offers advantages such as higher SNR, the employment of relatively inexpensive electronic sampling circuitry and operating with frequency modulated cw lasers. However, this all comes at costs of measurement time and the prerequisite on the whole system (including the fiber under test) to be linear and time invariant [74, 179]. Hence, the environmental impacts on the fiber under test are assumed to be stationary during the measurement. Furthermore, the iFFT only considers steady state vibrations only, which can lead to overshooting at sharp rising and falling edges known as the Gibbs phenomenon [181]. The overshooting issue was solved by implementing an iterative correction algorithm to minimize artifacts associated with acoustic wave relaxation [182]. This algorithm enabled the demonstration of spatial resolutions down to mm-scale [15].

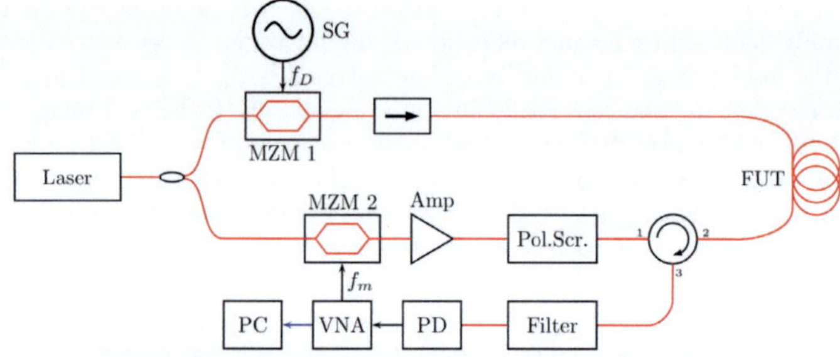

Figure 5.1: Fundamental BOFDA setup. Laser: Laser source, MZM: Mach-Zehnder modulator, Amp: amplifier, Pol.Scr.: polarization scrambler, FUT: fiber under test, Filter: optical filter, PD: photo diode, VNA: vector network analyzer, PC: personal computer

Figure 5.1 illustrates the basic BOFDA setup. In the pump path the spatial localization along the fiber is implemented. To obtain the complex transfer function, the modulation frequency f_m is successively swept in M equidistant frequency steps Δf_m. All applied frequency differences between pump and probe path f_D are achieved by the signal generator which drives the MZM1 in Fig. 5.1. In the time domain the discrete pulse response $g(t_q, f_D)$ is given using the inverse Fourier transform in Eqn. 5.3 [183]. The spatial pulse response $g(z_q, f_D)$ is obtained following the substitution shown in Eqn. 5.2.

$$g(t_q, f_D) = \frac{1}{M} \sum_{q=0}^{M-1} G(f_m, f_D) \, exp(j2\pi \, f_m t_q) \tag{5.3}$$
$$\text{with } t_q = \frac{q}{M\,\Delta f_m}; \; q = 0, 1, 2, ... M - 1$$

The BOFDA gain method is based on the modulation of the pump path. Thus, the complex baseband transfer function $G(f_m, f_D)$ is defined as given in Eqn. 5.4. $\hat{P}_P$ refers to the alternating part of the transmitted modulated pump power with its signal phase Ψ_P. In analogy the terms related to the the Stokes wave are indexed by S.

$$G(f_m, f_D) = \frac{\hat{P}_S(f_m, f_D)}{\hat{P}_P(f_m)} \; exp\left(j\Psi_S(f_m, f_D) - j\Psi_P(f_m)\right) \tag{5.4}$$

The pump and the Stokes wave in BOFDA are need to be considered as a function of distance due to the propagation loss of the fiber under test. At the fiber facet ($z = 0$), the pump and Stokes wave are given as [183]:

$$P_P(t, f_m)|_{z=0} = \bar{P}_P + \hat{P}_P(f_m) \; cos(2\pi f_m t + \Psi_P(f_m)) \tag{5.5}$$
$$P_S(t, f_m, f_D)|_{z=0} = \bar{P}_S(f_m, f_D) + \hat{P}_S(f_m, f_D) \; cos(2\pi f_m t + \Psi_S(f_m, f_D)) \tag{5.6}$$

The AC-coupling of the employed photo diode avoids measuring the stationary powers of pump and Stokes wave, $\bar{P}_P$ and $\bar{P}_S$ respectively [180]. Thus, equation (5.4) is valid for recording the transfer function. The dependence of modulation frequency in Eqn. 5.5 originates from the frequency-specific behavior of employed hardware components [180]. The alternating power of the Stokes wave $\hat{P}_S(f_m, f_D)$ is sensitive to the modulation frequency due to interference effects caused by the superposition of f_m with the Brillouin linewidth of the fiber test. This interference results in a low-pass characteristic specific for each fiber under test limiting the spatial resolution [180]. Though, a correction algorithm considering the acoustic wave relaxation was presented to overcome this limitation [182]. The phase term $\Psi_S(f_m, f_D)$ is caused by phase delay times of modulation components along the fiber length [180].

In order to apply the Fourier transformation in Eqn. 5.3, the prerequisite of the system linearity has be fulfilled. As shown in chapter 2, the pump wave and the Stokes wave are cross-connected via an induced acoustic wave. Furthermore, the relation between pump power P_P coupled into the fiber at $z = 0$ and Stokes power coupled into the fiber at $z = L$ can be simplified as exponential:

$$P_S|_{z=0} \sim P_S|_{z=L} \; exp(g \cdot P_P|_{z=0}) \tag{5.7}$$

Though, the exponential relation can be considered quasi-linear based in the following consideration. Inserting Eqn. 5.6 into Eqn. 5.7 and expressing the outcome as a second-order Tayler polynomial approximation gives [180]:

$$P_S|_{z=0} \sim P_S|_{z=L} \left[1 + g\bar{P}_P m_P \; cos(2\pi f_m t) + \frac{1}{4}\left(g\bar{P}_P m_P\right)^2 + \frac{1}{4}\left(g\bar{P}_P m_P\right)^2 \; cos(2\pi 2 f_m t)\right] \tag{5.8}$$

Equation 5.8 demonstrates, that the generation of harmonic waves can be suppressed by low pump powers $P_P|_{z=0}$ and by its low degree of modulation $m_P = \hat{P}_P/\bar{P}_P$. A value of m_P less than 1 can be achieved by an appropriate selection of VNA output power for f_m.

As for BOCDA, the spatial resolution and measurement range are linked by the modulation frequency in BOFDA. Due to the inverse Fourier transformation of discrete frequencies, a periodic and continuous pulse response $g(t)$ is obtained. In order to avoid multiple pulse equivalents along the sensing fiber, the maximum fiber length L_{max} is limited to [180,183]:

$$L_{max} = \frac{c}{2n \; \Delta f_m} = \frac{M \; c}{2n(f_m^{max} - f_m^{min})} \tag{5.9}$$

Due to the frequency-limited measurement of $G(f_m, f_D)$, the underlaying tranfer function of the complete system $G'(j\omega)$ is considered to be recorded over an infinite frequency range with a normalized rectangular function $rect(f_m)$ starting and stopping at f_m^{min} and f_m^{max}, respectively. Consequently, the process is given as a convolution in the time-domain [180]:

$$g(t) = g'(t) * \mathcal{F}^{-1} \left\{ rect(f_m) \right\} \tag{5.10}$$

Assuming $f_m^{max} \gg f_m^{min}$ and $f_m^{min} \to 0$, Eqn. 5.10 simplifies to:

$$g(t) = f_m^{max} \cdot g'(t) * sinc(2\pi f_m^{max} t) \tag{5.11}$$

The width of the sinc main peak described in Eqn. 5.11 is predominantly affected by f_m^{max} and directly relates to the spatial resolution Δz in BOFDA [179, 180, 183]:

$$\Delta z = \frac{c}{2n \; f_m^{max}} \tag{5.12}$$

However, the key limitation of the spatial resolution Δz in BOFDA is caused by the low-pass characteristic of the modulated optical signals depending the Brillouin linewidth of the fiber under test as mentioned earlier. The low-pass characteristic manifests itself as a degradation of SNR in the recorded BGSs. Taking into account the acoustic wave relaxation in a correction algorithm [182] and a high-pass filtering for accurate BGS reconstruction [184], spatial resolutions down to mm-scale were achieved [15]. Furthermore, in order to avoid power influences originating from the sinc-function sides lobes and to increase measurement accuracy, the measured $G(f_m, f_D)$ is low-pass filtered before applying the iFFT. Reference [180] has given a closer look on this topic. For all distributed measurements in this thesis and in [13, 185] a Kaiser window was used with a Beta-Factor of 5 [186].

The most important objection for the application of BOFDA is the required measurement time. There are field application with temporal dynamics in the range of the measurement time of a suitable BOFDA system. However, these mentioned time scales contradict the prerequisite of time invariance for the iFFT and therefore the application of a BOFDA system. All BOFDA systems resolve distributed BGSs by sweeping f_m from f_m^{min} to f_m^{max} in M steps for each f_D in f_D^{min} to f_D^{max}. The frequency resolution δ of each obtained spatial BGS corresponds to the frequency increment within two steps of f_D. The measurement time t_{aq} required for recording **one** $G(f_m, f_D)$ trace on a VNA is [65]:

$$t_{aq} = \frac{3L}{\Delta z} \frac{1}{RBW} = \frac{3M}{RBW} \tag{5.13}$$

The resolution bandwidth of the VNA RBW is need to be chosen with respect to the sampling frequency of the polarization scrambler f_{pol}. Apart of polarization maintaining SMFs, all fibers under test are randomly changing the state of polarization (SOP) along its length due to birefringence. In order to ensure an interaction between the counter-propagating pump and probe along the complete fiber length, a polarization scrambler is employed in a basic BOFDA setup (see Fig. 5.1). Consequently, insensitive fiber parts due to orthogonal pump and probe waves were avoided, which is known as polarization fading. To avoid polarization fading in $G(f_m, f_D)$ measurements, f_{pol} is need to be significantly greater than RBW, which leads to an averaging over several SOPs. As a rule of thumb the relation $f_{pol} \geq 8...10 \cdot RBW$ was given in [74, 97, 183]. Hence, the polarization scrambler limits the minimal measurement time of BOFDA systems. Employing polarization maintaining fiber as sensors drastically decreases the required measurement time as no polarization scrambler is needed. The measurement accuracy of BOFDA is improved with decreasing δ values, increased averaging samples of $G(f_m, f_D)$ and decreased RBW values. These three parameters directly affect the measurement time. Thus, BOFDA measurement times are a compromise between sensor length, sensing accuracy and spatial resolution. Though, a VNA is clearly over-designed for simple S_{21} measurements (see Eqn. 5.4). In a concept of a digital BOFDA [74] the complete data processing can be performed off-line with significant data acquisition time reductions.

Based on BOFDA numerous scientific publications have been presented, which can be divided in two major branches. First, based on the correction algorithm [182] and a high-pass filtering [184], spatial resolutions down to mm-scale were achieved [15, 187]. Second, long range BOFDA systems based on concatenating fibers with slightly different Brillouin shifts have been presented up to 63 km in a 100 km fiber loop without using distributed Raman amplification or image processing methods [185]. BOFDA assisted by bi-directional first-order Raman amplification and digital high-pass filters was presented in a 200 km fiber loop [13]. A single-ended variant L-BOFDA was proposed in [188], where the Fresnel reflection from the far end of the fiber is utilized to reflect the probe wave. Another single-ended variant based on spontaneous Brillouin backscattering was presented [65] and demonstrated [189] using a reflectometry approach in the frequency-domain. A digital envelope detection was presented to increase the spatial resolution and the spectral resolution of in BOFDR [190].

5.4 Brillouin sensing in POF

Brillouin scattering has been investigated in polymer optical fibers such as PMMA-based POF, partially chlorinated graded-index POF and perfluorinated graded-index POF. In PMMA-based POF with a core diameter of 980 μm, the experimental results estimated the BFS around 13 GHz [191] and a temperature coefficient of -17 MHz/K at 650 nm, which is -14.5 times larger compared to silica fibers at 1550 nm [192]. However, no linear relation between BFS and strain was observed [192]. The BFS decreases monotonically but not linearly with the increasing of environmental humidity and temperature [193]. In partially chlorinated graded-index POF the BFS was measured at 4.43 GHz at a wavelength of 1550 nm [194]. The temperature coefficient was determined at -6.9 MHz/K, nearly 5.8 times as large as the equivalent in silica fibers [194]. However, there is a drawback from an optical SBS sensing perspective. The BFS in both aforementioned POFs has been estimated utilizing external ultrasonic pulse-echos and without an optical stimulus of acoustic wave generation.

Brillouin interactions as know from silica fibers with an all optical stimulus has only been observed in PFGI-POFs, yet. In 2010 spontaneous [45] and in 2011 stimulated Brillouin backscattering [47] was presented in PFGI-POF. The BFS was measured around 2.8 GHz at a wavelength of 1550 nm [45]. The Brillouin threshold was observed to increase with increasing core diameters [195]. In contrast to silica fibers, the temperature and strain coefficients of the BFS are need to be given in ranges. For temperatures the BFS coefficient was determined to be C_T = -4.09 MHz/K in a range from 30 °C to 80 °C [35]. Later a wider temperature range from -160 °C to 125 °C was investigated and monotonically deceasing of the BFS was observed at different slopes. Therefore, C_T was determined to be -3.2 MHz/K below 60 °C and gradually changing up to -13.8 MHz/K within a range from 80 °C to 125 °C [196]. The BFS strain coefficent was determined to be C_S = -121.8 MHz/% at 1550 nm and up to 1 % of tensile strain, which is only -3.5 times as large as that in silica fibers [35]. Later tensile strain up to 60 % was investigated. A slim-down process of fiber and core diameter ending in a plastic deformed fiber was described [197]. The PFGI-POF strained up until 2.3 % was observed to correspond to the earlier determined strain coefficient of C_S = -121.8 MHz/% [197]. In a range between 2.3% and 10% of tensile strain, the BFS abruptly shifts to a higher frequency (approximately 3.2 GHz), called frequency hopping. Strain ranges above 10 % and up to 60% demonstrate no BFS dependency on strain [197]. The PFGI-POF was observed to slim down in a stepwise manner and with increasing strain, the slimmed sections grew longer [197]. This phenomenon is assumed to be caused by the yielding of the overcladding layer made up of polycarbonate, and not because of the core or cladding layers [197]. Utilizing a plastic deformed fiber after tensile straining above 5% provides wider usable dynamic strain range [198]. However, the strain coefficent changes drastically from -76.5 MHz/% (0-1.15 %) over a strain insensitive range (1.2-3 %) to 16.6 MHz/% (3-30 %). Furthermore, the cross-sensitivities between temperature and strain have been presented in given ranges in [199]. The origin of cross sensitivities and the non-linear temperature behavior is largely attributed to the elastic modulus of the PFGI-POF to be a function of temperature [199]. Table 5.1 summarizes all BFS coefficients in PFGI-POF.

Table 5.1: Brief summary of fiber coefficients describing SBS in PFGI-POF at a wavelength of 1550 nm

Fiber coefficent	Value
BFS	$\approx$2.83 GHz at room temperature [35]
Brillouin linewidth	$\approx$120 MHz at room temperature [35]
Temperature coefficient C_T	-4.09 MHz/K (30 °C to 80 °C [35])
	-3.2 MHz/K (-160 °C to 60°C [196])
	-3.2 MHz/K to -13.8 MHz/K (80 °C to 125 °C [196])
Strain coefficent C_S	-121.8 MHz/% (0 - 2.3 % [197])
	> 2.3 %: frequency hopping + non-linear behavior [197]
	16.6 MHz/% (3 - 30 %) for pre-strained fibers [198]
Strain coefficient dependence on temperature	1.5 MHz/(% °C) (strain: 0 - 1.2 %) [199]
	-0.3 MHz/(% °C) (strain: 4.0 - 9.0 %) [199]
Temperature coefficient dependence on strain	1.5 MHz/(°C %) (strain: 0 - 1.2 %) [199]
	-0.3 MHz/(°C %) (strain: 4.0 - 9.0 %) [199]
	Independent (strain: >13 %) [199]

Although the BOTDA has the potential to resolve a spatially distributed BFS in PFGI-POF [200], the high power optical pulses injected into the fiber potentially burn the SMF/PFGI-POF interface [201] or even cause a fiber fuse [202]. High optical pulse powers are needed to exceed the Brillouin threshold in PFGI-POF due to the high propagation loss and consequently short fiber length (see chapter 3). Instead, non-pulsed distributed SBS sensing approaches have been investigated. BOCDR in PFGI-POF was intensively studied in the last years [48–50]. Nowadays, spatial resolutions in the centimeter-scale are achieved with slope assisted BOCDR [51]. The first and only demonstration of BOFDA in PFGI-POFs so far was presented in [18] characterized by 4 m spatial resolution in a 20 m POF fiber with temperature change. Furthermore, strong temperature dependency of SOF/PFGI-POF interconnection utilizing FC/PC connectors has been described [18]. However, all presented distributed Brillouin sensors in PFGI-POF have not achieved sensing ranges over 25 m, due to the wavelength of operation at 1550 nm, which corresponds to an extremely high propagation loss of approximately 250 dB/km.

CHAPTER 6

Sensory analysis of SBS in PFGI-POF

Leon Brillouin described the temperature dependency of the BFS in 1922 [53]. The Stokes wave amplification in stimulated Brillouin backscattering is caused by light fraction on longitudinal acoustic waves oscillating in radial direction of a fiber. According to Eqn. 2.25, the Brillouin frequency shift is a function of the acoustic wave properties. The speed of this acoustic wave is a function of the materials density and elastic modulus. Thus, the BFS is a function of temperature and strain. In optical fibers the linear relation of the BFS to temperature and tensile strain have been reported as a BFS change coefficent for SMF and doped fibers [11, 203–206]. These coefficients describing the BFS change respectively to its corresponding environmental influence, enable an analytic expression of SBS-based sensors.

In this chapter the influence of temperature, strain and humidity on the BGS are investigated. The Brillouin frequency shift coefficients are determined for Brillouin PFGI-POF sensors at 1319 nm and are used to analytically describe the impact of temperature, strain and humidity. Combining the obtained measurement results a humidity-induced strain caused by a swelling of the coating was identified as the dominant origin of the humidity-induced Brillouin frequency shift.

Large parts of the following text in this chapter have been published in [21, 22, 207].

6.1 Measurement setup

Temperature and humidity

Fig. 6.1 displays the experimental setup. It mainly follows the high resolution, self-heterodyne detection of [48, 208]. The used neodymium-doped yttrium aluminium garnet (Nd:YAG) laser has a linewidth of 5 kHz at a wavelength of 1319 nm. Due to these beneficial characteristics, the Nd:YAG laser has demonstrated higher Brillouin scattering power in PFGI-POF compared to a 1550 nm (distributed feedback) DFB laser diode [46].

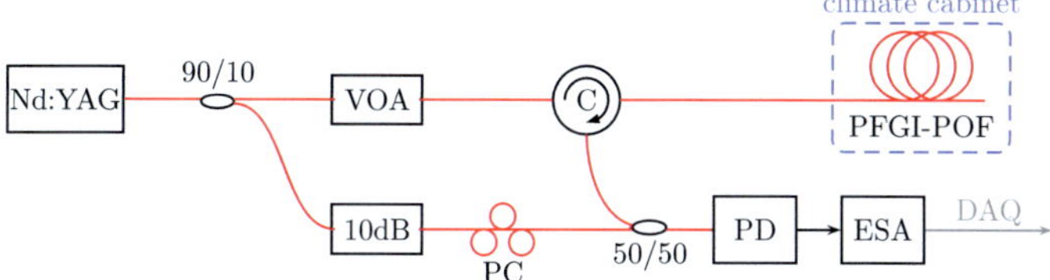

Figure 6.1: Experimental setup for investigating the influence of humidity on BFS. Nd:YAG: solid state laser at 1319 nm, VOA: variable optical attenuator, C: circulator, PFGI-POF: 200 m GigaPOF-50SR® by Chromis Fiberoptics, 10dB: 10 dB fixed attenuator, PC: polarization controller, PD: photo diode, ESA: electrical spectrum analyser, DAQ: data acquisition. [21]

The laser output power of 22 dBm was split into two paths with a ratio of 90/10 (pump path / local oscillator path). The less powerful path was used as a local oscillator for the heterodyne detection. Therefore, it was additionally attenuated by 10 dB and polarization controlled to maximize the SBS power level in each measurement. The constant input power at the combining coupler was 2.6 dBm.

The pump path power was adjusted by a variable optical attenuator (VOA) and injected into a 200 m PFGI-POF (GigaPOF-50SR® by Chromis Fiberoptics) after passing the circulator. The optical beat signal of backscattered Stokes light from the PFGI-POF and local oscillator path was detected by a photo diode (PD). The beat spectrum was measured by an electrical spectrum analyzer (ESA). All measurements were performed at a resolution bandwidth of 300 kHz, a spectral resolution of 1.6 MHz, 1001 spectral points and an averaging of 1000 samples.

All optical paths except the POF were SMF and coupled via FC/APC connectors, including the PFGI-POF which featured an FC/APC on both ends. The connectors guarantee accurate coupling into the fundamental mode of the PFGI-POF, to maximize the SBS power level [108, 111]. In contrast to [195] a shorter wavelength was applied, but we assume most stimulated backscattering power is generated on more than the first tens of meter of the fiber. Due to missing information on the mode coupling at 1319 nm in PFGI-POF, the chosen length of the PFGI-POF was 200 m to maximize the SBS power level.

The total loss along the PFGI-POF at 1310 nm was measured to be 6.5 dB, which is consistent with reported propagation loss for this wavelength ($\approx$ 35 dB/km [52]). The insertion loss measured at the SMF-POF and POF-SMF interfaces are 0.14 dB and 7.67 dB, respectively.

For all humidity and temperature measurements the PFGI-POF was loosely placed on a 1 cm thick polyvinylchlorid (PVC) plate to prevent strain induced BFSs. A longterm fiber annealing for 72 hours at 70 °C and 95 % relative humidity (r.h.) was performed in order to remove residual stress and avoid fiber shrinkage during the later performed measurements [92, 209].

After annealing, all BGS recordings were performed at constant temperatures under variation of the relative humidity. Within each cycle the relative humidity was increased and subsequently decreased from 20 %r.h. to 95 %r.h. within the same steps. The diffusion of water into polymers is a relatively slow entropy-driven process dependent on the relative humidity and temperature of the environment [92]. In order to achieve a homogeneous water distribution within the fiber, all recordings of the BGS were performed after 12 hours of exposure time in between each step. All changes in humidity and temperature were performed utilizing a Vötsch VCL4006 climate cabinet with a maximum measurement uncertainty in temperature and relative humidity over time of $\Delta T_{max} = \pm\, 0.5$ K and $\Delta h_{max} = \pm\, 3$ %r.h., respectively. Within a two-minute interval, used for all ESA measurements, both uncertainties are determined to be uniformly distributed and quantified to be $\Delta T = \pm\, 0.05$ K and $\Delta h = \pm\, 0.1$ %.

The BGS at each humidity step was averaged from 10 single measurements (each averaging 1000 samples), in order to minimize the influence of the climate cabinet's temperature uncertainty on the extracted BFS. From each mean BGS point its corresponding noise level was subtracted. The noise was dominated by relative intensity noise, due to the high power of the local oscillator path. Subsequently, a Lorentzian regression function was computed to determine the BFS (see next section).

Strain

Using the aforementioned experimental setup the BFS was measured while applying uniform strain to the first 90 m of a 200 m PFGI-POF. The fiber was strained using two fiber spools, whereby one was static and the second one was mounted on a stepper motor, controlling the distance between both spools. The applied strain was considered uniform since the spools were mounted to allow a rotation around their middle axis. The spools were simultaneously rotated half a turn before each measurement to release tension inhomogeneities. The first 90 m of the fiber were wrapped between the spools whereas the remaining fiber was freely placed to prevent further strain-induced BFSs. This fiber segmentation allows a temperature-compensated determination of the BFS for all strain values greater than 0.6 %. Figure 6.2 shows two measured BGSs with applied strain values of 1.2 % and 1.5 %.

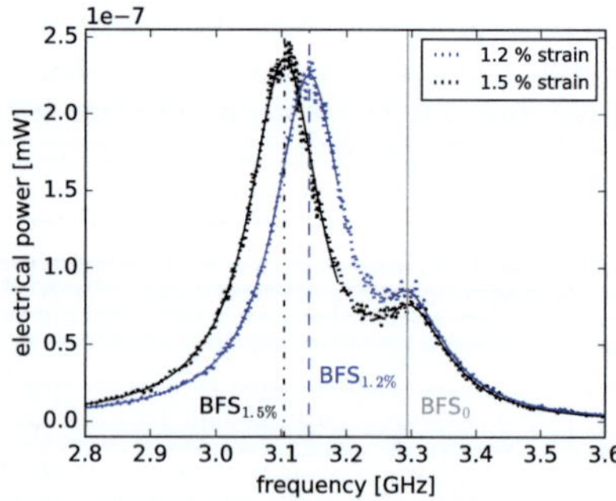

Figure 6.2:   BGSs of a 200 m PFGI-POF with applied strain values of 1.2 % and 1.5 % in the first 90 m, measured by an ESA. The regression functions of the strained and non-strain fiber segments are shown as well as their corresponding BFS. [22]

The BGS relating to the first 90 m fiber segment is affected by temperature fluctuations and strain changes, whereas the BGS of the non-strained segment is dominantly influenced by temperature fluctuations. The BGSs of the two fiber segments are separating from each other, since the applied strain causes a greater BFS change compared to the BFS change caused by temperature fluctuations. Thus, two Lorentzian regression functions were computed for each measured electrical spectrum to determine the BFS corresponding to each fiber segment (see BFS_0, $BFS_{1.2\%}$ and $BFS_{1.5\%}$ in Fig. 6.2). The strain-induced BFS change is determined by subtracting the BFS of the strained section from the BFS of the unstrained section (for example $BFS_{1.2\%}$ - BFS_0).

However, the spectral separation of the two BGSs was not feasible for strain values below 0.6 %, as the two BGSs are superposed. Therefore, a Lorentzian regression calculation was applied to the complete measured spectrum. The determination of strain-related BFS using the complete spectrum was possible since the SBS power of the strained section was dominant compared to the unstrained section. Unfortunately, the obtained BFS contains strain and temperature information which increases the uncertainty of the measurements compared to the BGS separation approach. The strain-induced BFS change was determined by subtracting the BFS of the strained fiber from the BFS of the unstrained fiber. All strain measurements were performed at room temperature. The maximum temperature drift during a single measurement was observed to be $\pm$ 0.5 K. The BGS at each applied strain level was averaged from 1000 samples.

6.2 Determination of the BFS

As described in chapter 2 the BFS is a function of strain and temperature. Therefore, a accurate determination of the BFS from the BGS is key. In order to determine trustworthy statistical BFS values even in between the BGS measurement points, Lorentzian [185] and parabolic [210] regression functions are used. Since, all regression calculations are based on optimization, accurate optimization algorithms are essential.

The Gauss-Newton method describes the first optimization algorithm characterized by fast convergence to the closest minimum. However, depending on the curvature of the target function and the accuracy of the initial parameter vector, the Gauss-Newton method tends to stuck in saddle points and can walk into the wrong direction [211]. Whereas, the method of gradient-descent utilizes the expansion of the target function by a Taylor series with the use of the first derivatives only [211]. Therefore, this algorithm walks strictly downhill and provides highly accurate data, when the initial parameter vector is nearby the requested minimum.

The Levenberg-Marquardt algorithm (LMA) was first published by Kenneth Levenberg in 1944 [212] and further improved by Donald Marquardt in 1963 [213]. The LMA combines the advantages of the Gauss–Newton method and the method of gradient-descent by using a damping factor [211]. The damping factor modifies the LMA during the optimization from Gauss-Newton method to the method of gradient-descent. This procedure ensures an increased robustness of the initial parameter vector and a high accuracy of the estimated parameter set [211]. Hence, the LMA was used to determine all BFS values from measured BGSs in this work by running the Python script in listing 6.1.

Listing 6.1: Python 3.5 source code to determine BFS from a given BGS measurement file

```python
import numpy as np
from scipy.optimize import leastsq #Levenberg-Marquardt algorithm
from scipy import stats

def lorentzian (x,p):
"""returns a lorentzian function by giving a vector of x-values and a set
of parameters (HWHM, centerpeakposition,formfactor,offset(=noiselevel))"""
    numerator = (p[0]**2)
    denominator = ( x-(p[1]) )**2 + p[0]**2
    y = p[2]*(numerator/denominator) + p[3]
    return y

def residuals_lorentzian (p,y,x):
"""returns a residual vector for optimization"""
    err = y - lorentzian(x,p)
    return err

def rsquared (f1,f2):
"""returns the R^2 value of a fit. Requires the measurement datas and the
fit. Also: from scipy import stats"""
    slope, intercept, r_value, p_value, std_err = stats.linregress(f1,f2)
    return 100*r_value**2

def fitting_lorentzian (ms):
"""returns a set of optimal parameters and a lorentzian function as vector
of function values by inserting a two-columned array of measurement
data. Requires the functions "residuals_lorentzian", "lorenzian" and "rsquared".
Also: from scipy.optimize import leastsq"""
    pbest = leastsq(residuals_lorentzian,p,args=(ms[:,1],ms[:,0]),full_output=1)
    best_parameters = pbest[0] #Selecting the best set of parameters
    fit = lorentzian(ms[:,0],best_parameters) #calculating the fit values
    r2 = rsquared(ms[:,1],fit) #calculating the R^2 value
    #max_value = max(fit) # optional determination of the peak power value
    return best_parameters,fit,r2


p = [30, 11000, 20, 1e-10] #initial parameter vector for Lorentzian in SOF
para, fit, r2 = fitting_lorentzian(np.vstack((Frequency,measured_intensity)).T)
```

The usage of the declared functions is shown in line 38. The variable para contains the four
parameters best possibly describing the measurement data Frequency and measured_intensity as
a Lorentzian function. However, the initial parameter vector p given in line 37 needs to be
set and controlled to ensure an accurate determination of the BFS.

Next to the determination of the BFS, the usage of regression calculations enables the
reduction of noise of each measured BGS. In terms of distributed fiber sensors (see
chapter 5), increased fiber lengths can be handled for sensing applications [151,185,214,
215].

6.3 Pump threshold in PFGI-POF

The hand-crafted connectors described in chapter 4 were evaluated utilizing the self-
heterodyne detection setup. Due to the SMF-PFGI-POF coupling the pump power excites
predominantly the fundamental mode of the PFGI-POF, which has presumably the highest
Brillouin gain coefficient and lowest Brillouin threshold. In addition, the mismatch of
mode field diameters at the interconnection results in a modal selection. Only a portion of
light carried in the fundamental mode of the PFGI-POF is detected. Thus, the generated
stimulated-amplified spontaneous Brillouin scattering in PFGI-POF is assumed to be
predominantly evaluated for the fundamental mode. In order to confirm these assumptions,
the obtained BGS is decomposed into two acoustical modes by applying a regression

function composed of a superposition of two Lorentzian functions. Figure 6.3 shows the measured BGS and the two computed Lorentzian functions separately.

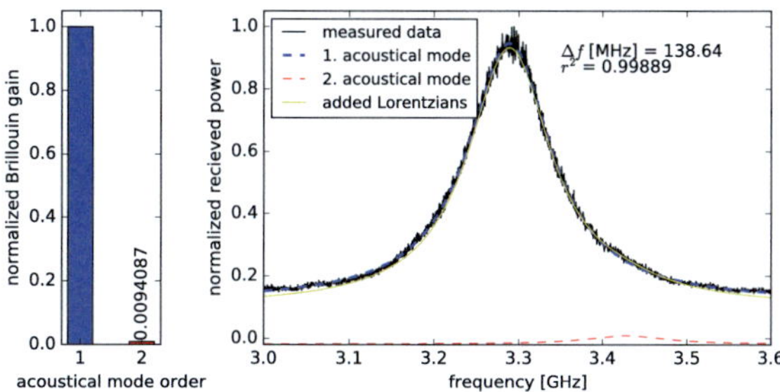

Figure 6.3: Decomposition of a measured BGS into two dominant acoustical modes.

It was found, that obtained BGS is almost perfectly symmetric and dominated by one acoustic mode. The second computed Lorentzian function is approximately 100 times weaker compared to the dominant acoustic mode. In addition, the second acoustical mode found is 138.6 MHz shifted towards higher frequencies, which is consistent to Eqn. 3.3 - Eqn. 3.6. Due to the negligibly small influence of the second acoustical mode, all BGSs measured with the hand-crafted interconnection and self-heterodyne detection setup have been evaluated by a single Lorentzian regression function.

However, a detailed modal analysis for SBS gain coefficients for different optical modes in PFGI-POF was not performed. Since the index profile of silica multimode fibers and PFGI-POFs are comparable, the SBS gain coefficient in PFGI-POF is assumed to comparatively follow its equivalent in silica fibers.

The Brillouin threshold in PFGI-POF is known to be function of fiber core diameter [45, 111, 195], fiber length (see chapter 2), fiber attenuation at the pump wavelength (see chapter 2) and the excited modal distribution (see chapter 2 and 4). The Brillouin gain tends to have a temperature influence is given for wavelengths nearby vibrational absorptions as the fiber loss increases in these wavelength regions. Only the most general definition of the threshold (section 2.3) can be applied for PFGI-POF due to the high fiber loss and therefore high pump power. The variation of pump power in the measurement setup provided the opportunity to determine the pump at which the backscattered Stokes power strongly increases. Figure 6.4 shows the obtained BGS at different pump powers. With increasing pump power, the Brillouin amplification gains and the backscattered Stokes wave increases, which is consistent to chapter 2. Furthermore, with increasing pump powers, the relative intensity noise at the photo diode was observed to rise the noise level. Consequently, the noise level was measured separately for each pump power to enable an evaluation of the BGS dependence on pump power.

The circulator is central component of the setup as it separates the incident and the scattered light of the fiber under test. Thus, the Brillouin threshold was measured with an SMF circulator and a graded-index multimode circulator with a core diameter of 50 μm, in order to find possible improvements of coupling properties between setup components. In addition, the patch cord connecting PFGI-POF and circulator as well as the 50/50 coupler

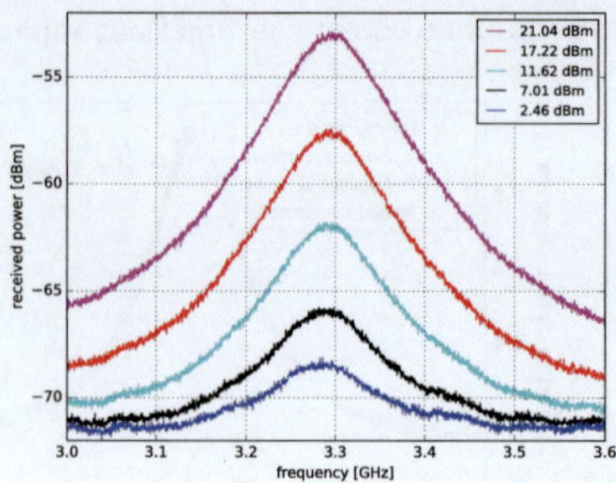

Figure 6.4: Obtained POF-BGS under variation of pump power by utilizing a SMF circulator.

were replaced with their multimodal equivalents. Each of the configurations exhibited exactly one critical MMF to SMF transition in the Stokes light path. Figure 6.5 shows the measurement results evaluated for the Brillouin linewidth and the BGS peak power over noise.

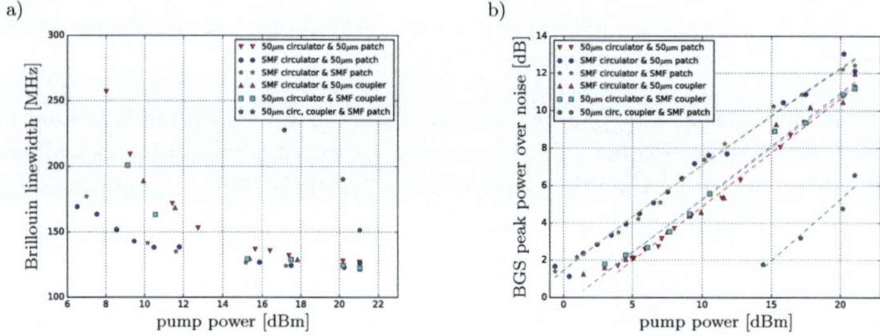

Figure 6.5: BGS dependence on pump power for given combinations of single-mode and multimode setup components. a) Brillouin linewidth as a function of pump power. b) Brillouin peak power over noise as a function of pump power.

For all varieties of SMF and MMF component combinations an increase of Stokes power with increasing pump powers was observed. Furthermore, all combinations demonstrate a linewidth narrowing with increasing pump powers which is consistent with [2,3,66]. All combinations can be summarized in three groups. First, the combination of all single-mode components and all single-mode with an multimode patch cord demonstrated comparable results obtaining the highest Stokes peak powers over noise and narrowest linewidths at lower pump powers, respectively. Second, the usage of multimode circulator and coupler with an single-mode patch cord. This combination marks the worst case scenario, as the modal selection takes place in the fiber illumination and evaluation. Therefore, the lowest Stokes peak power over noise and the broadest linewidths have been observed. Third, the usage of a multimode circulator without modal selection at the PFGI-POF. The results of this group are following the ones of the first group with a systematic offset. The reason for this offset is seen in the mode conversion of the multimode circulator. Figure 6.6 shows

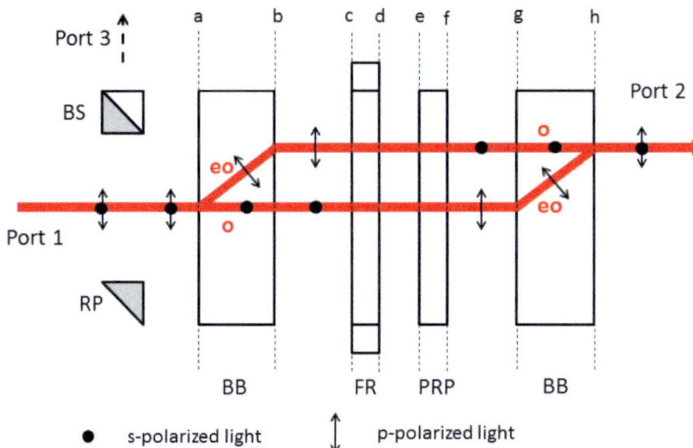

Figure 6.6: Schematic diagram of a 3-port circulator – one-way light propagation from port 1 to port 2. BB: birefringent block; FR: Faraday rotator; PRP: phase retardation plate; BS: beam splitter; RP: reflector prism o: ordinary ray; eo: extraordinary ray. Picture reworked from [216] and published in [207].

the typical light propagation in a 3-port optical circulator with light transmission from port 1 to port 2.

As presented by the light ray tracing marked in red, the light passes all circulators components in two rays of orthogonal polarization split at the interface of the first birefringent block cut at 45° to the optic axis. The light is finally recombined at the interface h of the second identical birefringent block [216]. In consequence of the orthogonal polarization states, no interference effects can occur in the circulator, even during the light recombining at the interface h. However, in case of the multimode circulators each interface a to h represents a refractive index perturbation [114,217] inducing mode coupling in the path of the pump signal. Moreover, due to the fact that unlike the ordinary ray o, the extraordinary ray eo does not satisfy Snell's law [218] the angles of refraction in birefringent media can be different for the two rays o and eo [219]. Additionally, considering the numerical aperture (NA) of MMFs the two orthogonal polarized rays have slightly different diameters in the section between interfaces b and g. This can lead to differing mode coupling in the two rays [207].

The loss mechanism of a multimode coupler is explainable by the fact that the two optical signals (Stokes and reference signal) generating the beat signal measured by the photo diode have different content of optical lower-order and higher-order modes. In other words, due to the design of a multimode coupler the process of coupling is characterized by a core-mode transmission of lower-ordered modes via one input port of the MMF coupler and by coupling over the higher-order modes via the second input port [207].

Thus, a single-mode circulator and coupler are used for all later distributed and integral measurements to avoid mode conversions and obtain higher Stokes powers over noise for spectral evaluations.

6.4 Temperature, humidity and strain influence on the BFS

Temperature and humidity

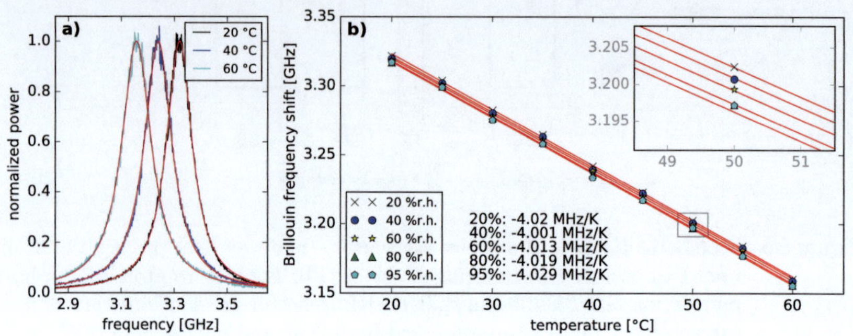

Figure 6.7: a) BGS at 60 %r.h. for given temperatures. b) Brillouin frequency shift over
temperature at specified relative humidities and their corresponding linear
regression functions. The inset provides a closer view of the data set around
50 °C. [21]

The BGS at each humidity step was averaged from 10 single measurements, in order to
minimize the BFS influence of the climate cabinet temperature uncertainty. From each
mean BGS point its corresponding noise level was subtracted, followed by a Lorentzian
regressions function computation to determine the BFS. Normalized example data and
their corresponding Lorentzian regression function at different temperatures are displayed
in Fig. 6.7a).

The BFS dependence on temperature and temperature BFS coefficient C_T for all measured
relative humidity steps is shown in Fig. 6.7b). A linear relation on temperature for fixed
relative humidities were observed. The temperature BFS coefficient for 1319 nm pump
wavelength is determined to be C_T = -4.02 MHz/K, which is close to C_T = -4.1 MHz/K
at 1550 nm [35]. Additionally, this coefficient exhibit no significant changes for variation
of relative humidities. Hence, the temperature BFS coefficient C_T can be considered to be
independent of humidity influences.

Even though the humidity does not influence the temperature BFS coefficient C_T, a BFS
offset between each linear regression function was observed. This humidity dependent
behavior of BGS in polymer optical fibers has not been reported yet. However, the origin
of the humidity induced BFS needs to be further investigated.

The BFS dependence on relative humidity at specified temperatures and their corresponding
relative humidity BFS coefficients $C_{h,rel}$ are displayed in Fig. 6.8a). Within each isothermal
step, a linear relation between BFS and relative humidity was observed as well as a smaller
BFS impact compared to temperature. The computed $C_{h,rel}$ values as a function of
temperature are plotted in Fig. 6.8b). This figure indicates $C_{h,rel}$ as a nonlinear function
of temperature.

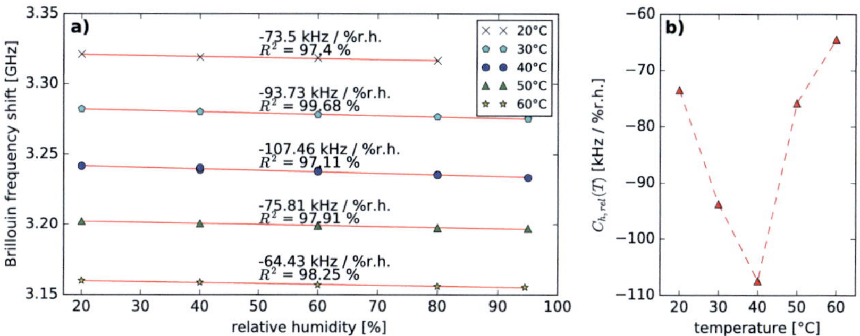

Figure 6.8: a) Brillouin frequency shift over relative humidities at selected temperatures and their corresponding isothermal linear regression function. b) Computed relative humidity BFS coefficient $C_{h,rel}(T)$ over temperature. [21]

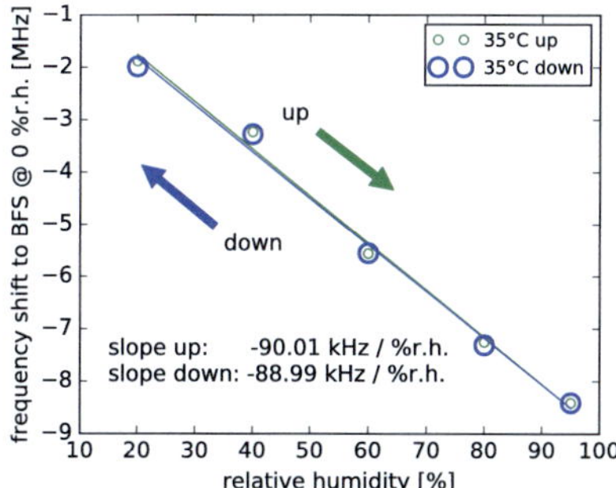

Figure 6.9: Brillouin frequency shift over relative humidities at 35 °C. Note that the frequency shift is given as a difference to the corresponding 0 % value, calculated for the linear regressions function as shown in Fig. 6.8b). [21]

The BFS hysteresis behavior under variation of humidity was studied more closely. Fig 6.9 displays all 9 measured isothermal humidity BFSs for 35 °C, separately. First, irrespective of the direction of the humidity change, the BFS exhibited a comparable slope of 89.5 kHz/%r.h. Second, no pair BFS measurement points at the same temperature and relative humidity was found to be greater than our measurement uncertainty of $\Delta BFS = \pm 201$ kHz ($\Delta BFS = \Delta T \cdot C_T$). Thus, the humidity hysteresis effect of PFGI-POF is negligible within our measurement limits.

Strain

The strain coefficient of the BFS was determined for a wavelength of $\lambda = 1319$ nm. The BFS was measured while applying uniform strain to a 90 m segment of a PFGI-POF. The temperature compensated BFS is plotted against applied strain in Fig. 6.10. A linear

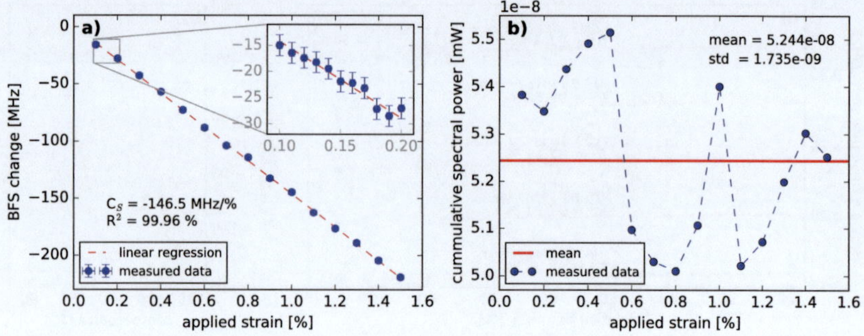

Figure 6.10:   a) Strain-induced change of the Brillouin frequency shift as a function of
strain and the calculated strain coefficient of the BFS in the range of 0.1 %
and 1.5 %. The inset provides a closer view on the measurement data in the
range of 0.1 % and 0.2 %. b) Cumulative BGS power of the whole measured
spectrum as a function of applied strain. [22]

relation between BFS change and applied strain was observed, which is consistent to [35]
within our measurement range. The strain coefficient of the BFS C_S was calculated to be
$C_S = (146.5 \pm 0.9)$ MHz/% in the range of 0.1 % and 1.5 % strain. The maximum applied
fiber strain in our experiment was chosen to be 1.5 %, since strains greater than 2 %
result in a non-linear BFS behavior [197]. Even greater applied strains (starting at 7.3%)
lead to an additional BGS at a higher frequency appears and this power increases with
increasing strain values, whereas the power of the original BGS decreases with increasing
strain values [197].

Figure 6.10b) shows the cumulative spectral power of the noise-subtracted BGSs within
a frequency span of 1.6 GHz plotted against applied strain. A non-linear course over
strain was observed. The power fluctuations are associated with laser and SBS power
fluctuations. The latter are the consequence of polarization effects. Thus, strain up to
1.5 % does not significantly affect the cumulative BGS power.

6.5 Analytical description of impacts

The BFS can be analytically described as follows:

$$BFS(T, h, \varepsilon) = C_T \cdot T + C_h(T) \cdot h + C_S \cdot S + BFS_0 \qquad (6.1)$$

BFS is the Brillouin frequency shift, h represents the humidity in % or g/m^3, T the
temperature in °C and S the strain in %. BFS_0 is the theoretical BFS at $T = 0\,°C$ and
$h = 0$ %r.h. in Hz for an unstrained fiber. The value of C_S was determined in the previous
section. For clarification it needs to be pointed out, that C_T and $C_h(T)$ will be determined
for unstrained PFGI-POFs. The BFS cross-effect of strain and temperature was discussed
in [199] and the cross-effect of strain and humidity should be further investigated in the
future. However, the dependence of $C_{h,rel}$ on temperature is a non-monotone function
(see Fig. 6.8b)). We converted the absolute humidity from the relative humidity values

of Fig. 6.8a) into absolute humidities by using the improved Magnus form [220], the molecular weight of water vapor (18.016 kg kmol^{-1} [221]) and the universal gas constant (8314.4 J kmol^{-1} K^{-1} [222]). The BFS is plotted against absolute humidity in Fig. 6.11a). In contrast to the relative humidity, the BFS coefficients over absolute humidity $C_{h,abs}$ can be described by a monotonous function in the whole measured temperature range. Furthermore, the applied linear regression functions also indicate a linear relation between BFS and humidity in each isothermal step.

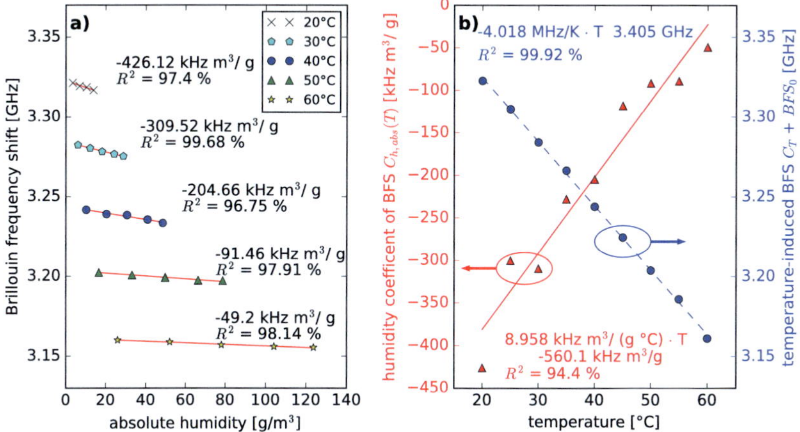

Figure 6.11: a) Brillouin frequency shift as a function of absolute humidity at selected temperatures. b) Absolute humidity and temperature-induced Brillouin frequency shifts at all measured temperatures. [21]

Taking into account the measurement data of Fig. 6.11a), the parameters C_T, $C_h(T)$ and BFS_0 can be calculated. A linear regression function was computed for each isothermal measurement step. The intercepts of those linear functions refer to a humidity value of $h = 0$ g/m^3. Therefore, the intercepts describe the temperature-induced BFS only, which are plotted in Fig. 6.11b) (see blue circles). C_T and BFS_0 refer to the parameters of a linear regression function of the intercept points (see blue dashed line as well as equations 6.2 and 6.5). In contrast, the slopes of the linear regression functions in Fig. 6.11a) describe the humidity coefficient of the BFS $C_h(T)$ (see red triangles in Fig. 6.11b)). The linear regression function of the slopes is given in Eqn. 6.3.

$$C_T \quad = \quad (-4.02 \pm 0.01) \text{ MHz/K} \tag{6.2}$$
$$C_h(T) = C_{h,abs}(T) \quad = \quad 8.96 \text{ kHz m}^3 \text{ / (g °C)} \cdot T - 560.1 \text{ kHz m}^3/\text{g} \tag{6.3}$$
$$C_S \quad = \quad (146.5 \pm 0.9) \text{ MHz/\%} \tag{6.4}$$
$$BFS_0 \quad = \quad (3404.8 \pm 0.5) \text{ MHz} \tag{6.5}$$

Thus, by using Eqn. 6.1 and the empirically determined values of $C_h(T)$, C_T, C_S and BFS_0, a BFS can be estimated for a given temperature, strain and absolute humidity value. Furthermore, the analytical description proves that temperature, humidity and strain influence the BFS independently. Hence, a clear distinction between temperature, humidity and strain is not possible by only measuring the BFS.

6.6 Influence of temperature and humidity on the BGS

On basis of the obtained noise-subtracted and non-distributed BGSs, the humidity-induced changes of the stimulated Brillouin backscattering power were analyzed. A cumulative SBS power level was determined by integrating over all spectral power levels of the BGS within a 1.6 GHz span around the center frequency. Figure 6.12a) displays the cumulative spectral power for different relative humidity settings. The SBS power decreases with increasing relative humidity and temperature.

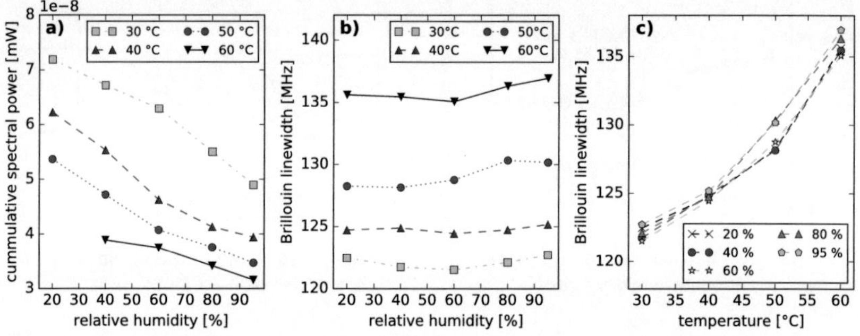

Figure 6.12:   **a)** Cumulative BGS power of the electrical signal as a function of relative humidity at specified temperatures. **b)** Brillouin linewidth (full width at half maximum) as function of relative humidity at given temperatures.

Figure 6.12b) and c) show the linewidth of the BGS from a 200 m long PFGI-POF as a function of relative humidity. This figure demonstrates no significant influence of humidity on the Brillouin linewidth. However, a sensitivity of the Brillouin linewidth on temperature was observed, which is consistent to [195]. The Brillouin linewidth in PFGI-POFs increases with increasing temperatures, whereas the Brillouin linewidth in silica fibers decreases with increasing temperatures in the same temperature range [223].

This phenomenon can be explained when taking into account the thermoplastic properties of both, CYTOP and polycarbonate [52,224]. The polymer chains are associated through intermolecular forces, which weaken with increasing temperatures [92,224]. Thus, an increased moldability and a reduced resistance for external deformation is achieved by heating [224] and can be considered as a reduced elastic modulus with rising temperatures. The temperature dependency of the elastic modulus in PFGI-POF was demonstrated in [93]. The cross effect of temperature and strain on the BFS presented in [199] can be considered as temperature-induced changes of the elastic modulus. Hence, an elevated temperature can be considered to lead to a reduced phonon lifetime and consequently to a greater Brillouin linewidth [1]. This spectral spreading of the BGS with risen temperatures causes a lowered Brillouin gain coefficient [1]. Consequently, the **temperature-induced**

reduction of the SBS power level can be considered to be dominantly caused by the thermoplastic characteristic of CYTOP and the vibrational absorption dependence (see figure 6.13b) and 6.14b)).

The origin of the **humidity-induced SBS power reduction** can be largely attributed to the vibrational absorption of the fiber core material. While increased humidity levels lead to higher fiber losses (see figure 6.13b) and 6.14b)), the effective length of the PFGI-POF decreases [1]. Consequently, the Brillouin pump threshold rises [1]. Therefore, we expect the SBS interaction to occur on a shorter fiber length. In addition, the Brillouin backscattered light is attenuated. However, humidity-induced changes of the Brillouin gain coefficient should be further investigated in the future.

6.7 Origin of the humidity-induced BFS

According to Eqn. 2.25 the humidity-induced BFS can be explained by changes in the core material parameters n, ϵ and ρ. The humidity variations affect these parameters directly due to water absorption into the fiber core or indirectly as a result of humidity-induced fiber strain as described in the previous section. To be able to discuss the humidity-induced fiber strain, a further fiber characterization is needed.

Fiber characterization

Spectral absorption

The optical transmission spectra of the PFGI-POF were measured at 4 temperature and 5 humidity values in a climate cabinet. A broadband light source (Yokogawa AQ4305) launched light into a 200 m GigaPOF-50SR by Chromis Fiberoptics, which was previously annealed for 72 hours at 70°C and 95 % r.h.. An overfilled launch condition was satisfied. The transmitted light was detected by an optical spectrum analyzer (OSA, Advantest Q8347). Figure 6.13a) displays selected transmission spectra at specified relative humidity values.

The absorption peaks correlate to vibrational absorption bands of the C-F bond (multiple peaks around 1130 nm [225]) and the O-H bond (center of vibration at 1383 nm [225]). The C-F bond is the characteristic bond of CYTOP [52]. Increased optical attenuations were observed for rising humidity values in both absorption bands. The optical attenuation increase with rising humidity values of the O-H peak could originate from the change of the total amount of water molecules within the fiber core material. A possible cause for the humidity-depending behavior of the carbon–fluorine bond is seen in its characteristic to be highly polarized as it attracts partially charged water molecules [226].

The wavelengths of $\lambda_1 = 1064$ nm and $\lambda_2 = 1319$ nm are marked in Fig. 6.13a). λ_1 was chosen since there is little effect by vibrational absorption. Semiconductor laser diodes at λ_1 [227, 228] might be feasible for sensing or communication applications. The humidity-induced BFS was reported at λ_2 [21]. The wavelength of 1319 nm will be used for all Brillouin related investigations in this paper. The corresponding transmission powers at λ_1 and λ_2 are displayed as a function of relative humidity in Fig. 6.13b). In contrast to the transmission spectra at λ_1, the transmission spectra at λ_2 show a strong dependence

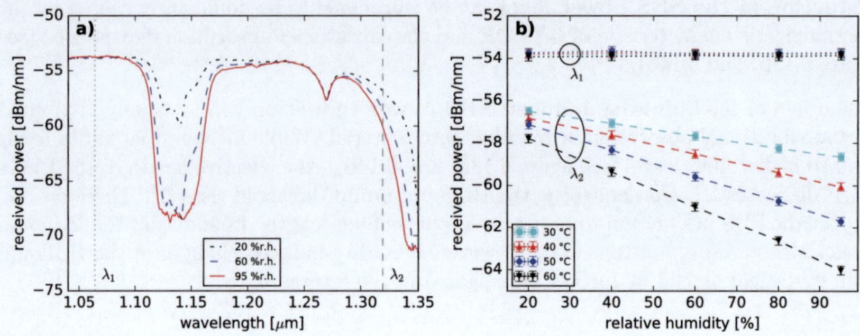

Figure 6.13:   **a)** Transmitted optical power spectra of a 200 m PFGI-POF under variation of relative humidity at 40 °C . **b)** Transmission power as a function of humidity at specified temperatures and two wavelengths ($\lambda_1 = 1064$ nm and $\lambda_2 = 1319$ nm). [22]

on relative humidity and temperature which is consistent with [94]. Increasing relative humidity leads to increased transmission losses as shown in [94]. Furthermore, the humidity sensitivity of λ_2 was observed as a function of temperature. Increased temperatures result in a greater transmission loss. This behavior originates from O-H and C-F bond vibrational resonances [229].

The measurement results are consistent to [94]. In addition, the presence of water inside the fiber core material is underlined and the humidity-driven absorption of water is demonstrated. However, the exact quantity of the water uptake into the core material is not determined.

Rayleigh backscattering

Swept-wavelength interferometry (Luna OBR 4413) was used to performed the following measurements. The Luna OBR4413 has a center wavelength of 1306 nm and the sweep range of 0.8 nm was set to achieve a spatial resolution of 5 cm. In contrast to the OSA measurements, a 1 m single mode silica fiber was used to connect the PFGI-POF to the Luna device. Consequently, the measured signals are dominated by the Rayleigh scattering of the fundamental mode, due to mode selective coupling. Nevertheless, the Rayleigh influence of higher optical modes on the measurement signal can not be excluded due to the strong mode coupling of the PFGI-POF [200]. Two example backscatter traces are shown in Fig. 6.14a).

Increasing humidity values were observed to form a sharper declining slope in the backscatter traces. A linear regression function was computed for each measured backscatter trace. The determined slope values are shown as a function of relative humidity for specified temperatures in Fig. 6.14b). As shown in Fig. 6.13b), the fiber attenuation increases with rising humidity and temperature values. Evidently, the fiber attenuation at 1306 nm is a function of temperature and humidity.

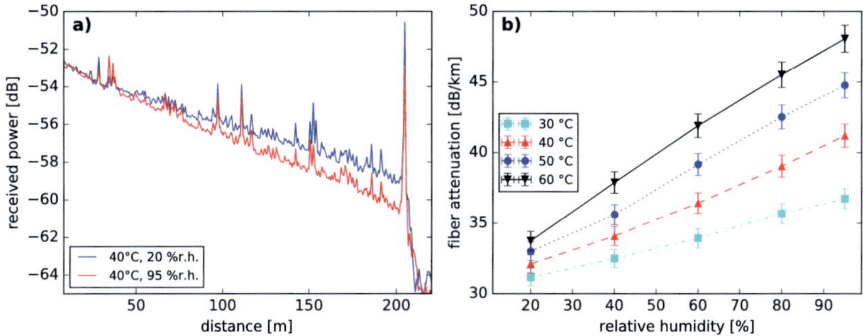

Figure 6.14: **a)** Measured Rayleigh backscatter trace at 40 °C and 20 %r.h., 95 %r.h., respectively. **b)** Fiber attenuation from backscatter power change as a function of relative humidity at given temperatures and a wavelength of 1306 nm. The error bars mark the threefold standard deviation from the linear regression applied on all measured data points. [22]

In addition to the variation in fiber attenuation, a fiber length change was observed in the backscatter traces. The measured fiber length changes are plotted in relation to the obtained fiber lengths at 20 %r.h. and 30°C in Fig. 6.15, respectively. Both data plots indicate a linear increase of measured fiber length for rising temperatures and humidities within the given measurement limits. There is no apparent cross-sensitivity of humidity and temperature change on the measured fiber length change. The computed slope values are considered as the coefficient of hygroscopic expansion (CHE) and coefficient of thermal expansion (CTE) for the tested fiber along the optical axis, respectively. The calculated expansion coefficient values are $CHE = (7.4 \pm 0.1) \cdot 10^{-6}$ %r.h.$^{-1}$ and $CTE = (22.7 \pm 0.3) \cdot 10^{-6}$ K^{-1} within the given measurement ranges. The given uncertainty values are the threefold standard deviation from the linear regression applied on all measured data points, respectively. On the basis of these coefficents, the measured fiber length l can be described as a function of temperature and humidity as presented in Eqn 6.6. l_0 refers to a fiber length at a given temperature and humidity. ΔT and Δh are the changes in temperature and humidity, respectively.

$$l(T, h) = l_0 \left(1 + \Delta T \cdot CTE + \Delta h \cdot CHE \right) \tag{6.6}$$

To calculate the spatially distributed backscatter trace a constant group index of $n = 1.3536$ [94] was used. However, the refractive index of CYTOP was reported to be a function of temperature ($dn/dT = $ -9.7$\cdot 10^{-5}$ K^{-1}, measured as a 3 μm thin film around 1550 nm [230]) and the absorbed water in the fiber core could cause changes in the group index. Thus, the measured fiber length should be considered as a superposition of refractive index change, temperature-induced expansion of the polymers [231] and humidity-induced swelling due to water intake [231]. Unfortunately, a distinction between the named effects is not possible using a time-of-flight-based measurement method. Therefore, Fig. 6.15 also displays the optical runtime change as a function of relative humidity and temperature. For clarification, all measured runtime changes are reported as fiber length changes.

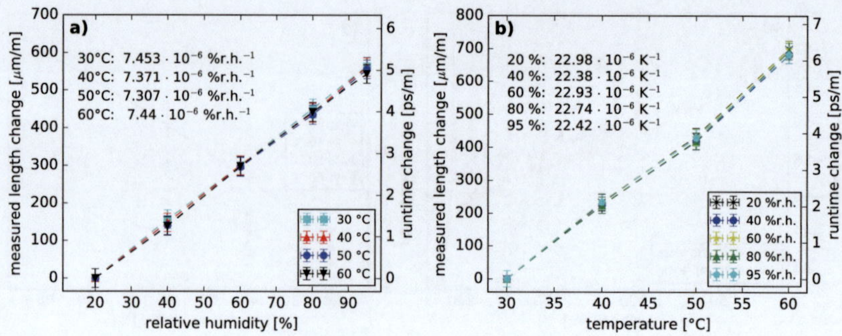

Figure 6.15:   Measured relative fiber length change and optical runtime changes induced by a) humidity as well as b) temperature. The displayed coefficients are the slopes of the linear regression functions computed for all temperature or humidity values based on the measured fiber length changes. [22]

The humidity-induced measured fiber length change may originate from the humidity-induced change of the fiber core properties and from the humidity-depending behavior of polycarbonate [231, 232], used as fiber coating. As to [231], the further the humidity of the fiber surrounding increases, the more water molecules are absorbed into the fiber as a result of elevated partial vapor pressures. Since, polycarbonate tends to absorb more water than CYTOP (0.04 %wt and <0.01 %wt, respectively [233]), the dominant impact is expected for the fiber coating. Consequently, a humidity-induced swelling of the polycarbonate is caused [231], resulting in an expansional straining of the fiber core along the optical axis. The described process is purely mechanical and independent on the measurement wavelength. The humidity sensitivity of the PFGI-POF strongly depends on the coating material and thickness. The measured humidity-induced fiber length change ($7.4\ \mu\varepsilon$ %r.h.$^{-1}$) is significant compared to polyimide and acrylate coated silica single mode fibers at standard layer thickness of 15 μm (1.29 $\mu\varepsilon$ %r.h.$^{-1}$ and 0.2 $\mu\varepsilon$ %r.h.$^{-1}$, respectively [19]). Though, silica fibers with a thicker polyimide coating demonstrate increased strain responses compared to the PFGI-POF (up to 38.5 $\mu\varepsilon$ %r.h.$^{-1}$ for 876 μm coating thickness [20]).

Origin of the humidity-induced BFS

As shown earlier, humidity and temperature are two mutually independent influences on the BFS. Thus, the temperature BFSs can be separated from the measured BFS to study the humidity-induced changes more closely. The BFS results of Fig. 6.7b) and 6.8a) are plotted as a function of relative humidity for different temperatures in Fig. 6.16.

At all temperatures a BFS decrease with rising humidity values was observed. As previously shown in section 6.7, humidity causes a measured fiber length change. Consequently, the coefficients $CHE = 7.4\cdot10^{-6}$ %r.h.$^{-1}$ and $C_S = $ -146.5 MHz/% were used to calculate a theoretical BFS coefficient describing the fiber strain provoked by humidity ($CHE \cdot C_S = $ (-108.4 $\pm$ 9.0) kHz/%r.h). Under the assumption that the measured length change due to humidity change can be predominantly contributed to the swelling of the fiber coating,

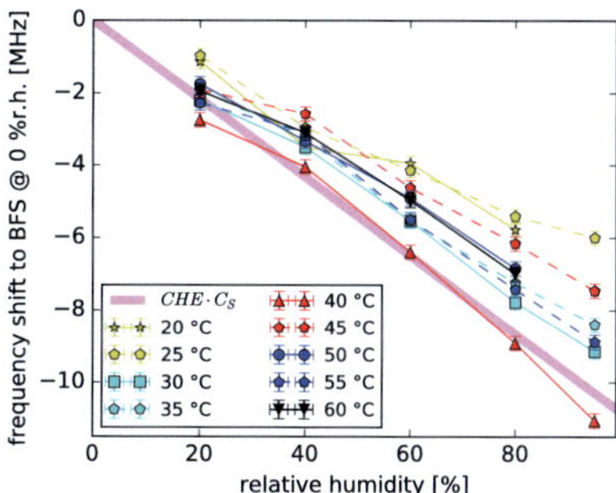

Figure 6.16: BFS as a function of relative humidity at specified temperatures. The
BFS is given as a difference to the theoretical BFS at each corresponding
temperature and 0 %r.h.. [22]

we plotted the swelling strain-equivalent dependence $CHE \cdot C_S$ in Fig. 6.16. As the
$CHE \cdot C_S$ and the measured BFSs are in good agreement, the humidity-induced strain
can be considered as the dominant cause of the humidity-induced BFS. In addition, the
theoretical coefficient $CHE \cdot C_S$ and therefore the BFS change caused by coating-induced
strain is independent on temperature influences within the measured temperature and
humidity ranges.

The isothermal BFS graphs demonstrate divergence from each other with rising humidity
values. However, a detailed analysis of the temperature behavior of the humidity-induced
core property changes apart of fiber strain can not be given based on this data set. The
coefficients C_S and CHE were determined at different wavelengths (1319 nm and 1306 nm,
respectively). These wavelengths are close to each other, though the exact influence of the
wavelength difference on the humidity-induced strain coefficient and its uncertainty should
be further investigated. Thus, a temperature dependence of the humidity-induced changes
of the core material parameters can not be excluded as it is indicated in the analytical
description for the humidity-induced BFS [21].

When taking into account the fiber strain and the change of the refractive index and/or
density, the lack of a humidity-induced BFS hysteresis can be explained. The humidity-
induced strain is within the fully elastic strain limit of the PFGI-POF (up to 2.5 % [94]).
Furthermore, the water absorption into the fiber can be considered fully reversible without
any damages for an annealed PFGI-POF as described for PMMA-POFs in [234].

CHAPTER 7

Frequency uncertainty in Brillouin sensors

Parts of the following text in this chapter are part of a paper under review whilst the thesis is in press.

A stochastic approach for the BFS uncertainty estimation is presented in order to extend the variety of underlaying mathematical regression functions. At first, the amplitude-noise in a SBS fiber sensor is shown to be affected by the measurement system and the underlying SBS process itself. The amplitude noises are demonstrated to follow the BGS across the measured spectral range for low numbers of averaging samples. Based on the this spectral noise behavior on the BGS, a Monte Carlo simulation for the synthesis of artificial noisy BGSs was performed. Differences in obtained BFSs of the artificial noisy BGSs are stochastically evaluated to estimate the maximum difference frequency within a 99.5 % confidence interval. This proposed approach enables the estimation of BFS uncertainties irrespective of employed mathematical regression functions, sensor fiber type and measurement method. The computing time of this procedure strongly depends on the chosen number of noise samples and standard deviations. On the basis of a distributed Brillouin fiber sensor the results of the stochastic approach demonstrated comparable outcomes to analytic results based on a parabolic regression function. An estimation of frequency uncertainties for Lorentzian regression function is presented on a distributed measurement. The analytic and stochastic approach can not provide a comprehensive description of BFS uncertainties for SBS-based fiber sensors. In field applications further influences are needed to be taken into account such as double peak structures in spectral responses or fiber effects causing additional BFSs. For comparability of SBS based distributed FOSs in the time domain, the figure-of-merit (FoM) was defined in [210]. The FoM includes next to the measurement setting and the BFS uncertainty of the FOS. For comparison purposes the stochastic noise analysis opens the door for FoMs irrespective of fiber types and measurement approach. The link between the FoM in time-domain and frequency-domain is shown.

Frequency determination using a Lorentzian curve fitting is the most common technique in several scientific fields. The estimation of the BFS from the measured BGS is of crucial importance for temperature and strain measurements based on SBS [1, 3, 11]. Any BFS uncertainties result in uncertainties of temperature or strain. All fiber optic sensors based on SBS have one similarity: The measured signals, which serve as a basis for a curve fitting, include different types of noises. In terms of distributed SBS-based fiber sensors, several sources of noise were identified such as laser phase noise [136], amplified spontaneous emissions of amplifiers [235], double Rayleigh scattering generated by the probe wave [235], noise induced by pump-probe interaction [235], quantization noise [236] as well as thermal and shot noise of the employed photodetector [236]. In addition, the characteristics of the fiber under test affects the accuracy of the extracted BFS as well. Exemplarily named are water- and humidity-induced strains employing a polymer optical fiber (POF) [21,22], a polyimide coated single-mode fiber (SMF) [19,20,237] or a acrylic coated SMF [19,238].

The noise sources and their influences in a Brillouin optical time domain analysis (BOTDA) setup are described in [210,235]. All presented amplitude noise-induced frequency uncertainties are directly proportional to a temperature or strain uncertainty. Apart from [239], which proposes a process to calculate the temperature uncertainty for distributed fiber sensors in general, there is no specific standardization of measurement uncertainties for SBS-based fiber optic sensors. Several approaches have been presented in which the frequency uncertainty is presented as a function of the signal-to-noise ratio [210] or quality factor [240,241]. All reported analytical descriptions [210,240,241] are based on parabolic regression functions to determine the BFS. In this chapter the amplitude noise and its dependence on the BGS is described and therefore a stochastic approach is proposed to estimate the frequency uncertainty. In contrast to the parabolic fit, the entire measured spectral response is taken into account to estimate the BFS uncertainty, due to the spectrally non-uniform amplitude noise distribution. In order to extend the definition of frequency uncertainty for SBS-based fiber sensors a generalized stochastic approach is described, enabling the determination of frequency uncertainties irrespective of employed methods for BFS determination such as Lorentzian regression functions [185], quadratic functions [210,241] or neuronal networks [242–244]. Furthermore, the presented approach provides the basis to compare and benchmark regression functions, implementation variants (Levenberg-Marquard [211,213], gradient-method [211], least-mean square [211] or by utilizing neuronal networks [211]) and employed computing hardware. In addition, the stochastic approach is valid irrespective of fiber type (SMF [185], polarization maintaining fibers [245], multi-mode fibers [246] or POF [21,22]), domain approach and setup configuration (BOTDA [11,132,246,247], BOTDR [11,248], BOFDA [11,13,15,185], BOFDR [65,190], BOCDA [11,162,249] or BOCDR [11,50,245]). It has to be noted, that slope-assisted SBS sensing methods are not discussed.

7.1 Description of the model

Most commonly the frequency uncertainty of Brillouin-based fiber sensors is estimated by the proposed procedure of [210]. Measurement parameters such as the Brillouin linewidth Δf_B, the scanning frequency step δ and the amplitude noise $\sigma^\star$ contribute significantly to the accuracy of the estimated BFS. The amplitude noise $\sigma^\star$ is defined as the standard deviation of a measured normalized power next to the BGS and considered as the system noise. Furthermore, the amplitude noise is expected to be uniform across the spectral range, which is the case for a high number of averaging samples. The amplitude noise is mainly affected by the setup components and their thermal noise and relative intensity noise. The standard approach of applying a parabolic fitting of all measured data above a given amplitude threshold η, provides an estimated BFS and its corresponding frequency uncertainty given as a standard deviation σ_v by following Eqn. 7.1 [210]. Equation 7.1 is valid if the frequency steps δ are significantly smaller than the Brillouin linewidth of the observed fiber.

$$\sigma_v = \sigma^\star \sqrt{\frac{3 \cdot \delta \cdot \Delta f_B}{8\sqrt{2}(1 - \eta)^{3/2}}} \xrightarrow{\eta=0.5} \sigma^\star \sqrt{\frac{3}{4}\delta \cdot \Delta f_B} \tag{7.1}$$

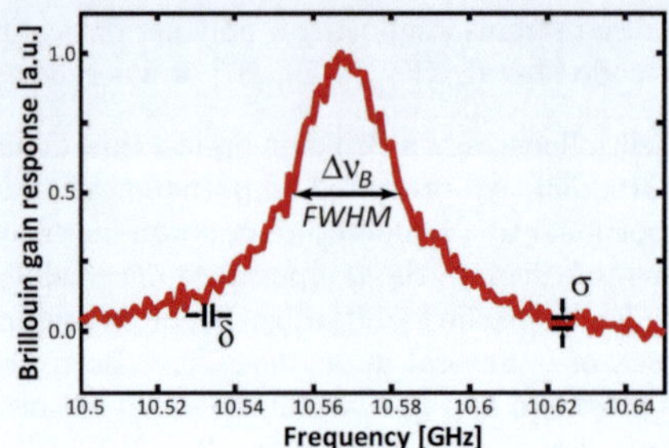

Figure 7.1: Amplitude noise on a spectral response induces uncertainties for the estimation of the peak gain frequency. Reprinted from [210], where the Brillouin linewidth is denoted by Δv_B.

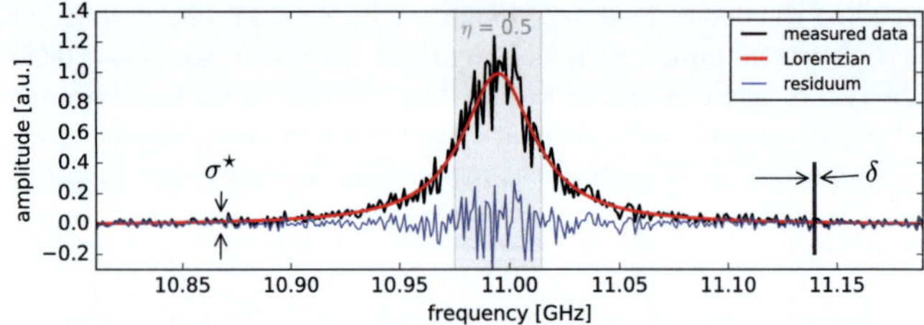

Figure 7.2: Measured spectral response of one spatial segment in a ultra low-loss SMF, measured by a BOTDA at a wavelength of 1550 nm with 100 averages in frequency steps of $\delta = 1$ MHz. The power levels of the BGS measurement were normalized to the amplitude of the determined Lorentzian regression function. The difference between measurement data and Lorentzian regression function is labeled residuum. The parameters used to estimate the frequency uncertainty following Eqn. 7.1 are illustrated.

Taking into account the theoretical description of the SBS-induced acoustic wave as a forced and damped oscillation [1,3], a Lorentzian function is most suited to approximate the measured spectral response of a SBS fiber sensor. Figure 7.2 shows a measured spectral response of one spatial segment and its corresponding Lorentzian regression function. In contrast to [210] applying a parabolic fitting only to a limited spectral range (indicated gray span in Fig. 7.2), the Lorentzian regression function covers the entire measured spectral range. The point-wise spectral difference between measurement data and the corresponding Lorentzian is considered to be a residuum. The residuum (see Fig. 7.2) is predominantly affected by the spectral behavior of the measurement system and by the underlaying SBS process itself. One observes, that the residuum is characterized by spectrally non-uniform noise amplitudes, which are more intense within the spectral range of the BGS. The cause of this spectral non-uniformity is need to be further investigated in the future. It can be observed, that the noise amplitudes and the BGS show a similar behavior in the spectral range. Investigating the residui of several mutually independent measurements shows that the average absolute values of the residuum exhibit a Lorentzian shape. Furthermore, the residuum is observed to have no mean value (see Fig. 7.2), but demonstrates random results for each single measurement. In virtue of these three characteristics (time-depended random process, lack of mean value, converges for squared values), the amplitude noise in SBS sensors can be assumed to be ergodic. However, this assumption might be valid from an observational point of view, but is need to be validated and/or contrasted with experimental data in future investigations. The validation of the proposed model with experimental would ensure that the resulting artificially-generated noisy Brillouin spectra are reliable and consistent with real systems. In general, this approach can be considered as a simplified Monte-Carlo simulation of [250], which presents the BGS as a complex function consisting of a real (nonlinear amplitude change) and an imaginary part (nonlinear phase shift).

Ergodic processes describe (quasi-)stationary processes with the help of covariances [251]. For the sake of simplicity and based on the pioneering work of [210], the amplitude noise σ is defined as the standard deviation of the residuum across the complete measured spectral range and is used in this paper to estimate the frequency uncertainty of SBS-based fiber sensors. In case of Fig. 7.2 the standard deviation of the entire residuum was determined to be $\sigma = 0.051186$. Following the proposed approach of [210] the standard deviation of noise is $\sigma^\star = 0.010601$ in the spectral range from 10.8 GHz to 10.9 GHz (101 spectral points). For clarification, the spectral scanning range and scanning steps δ of SBS-based sensing techniques strongly influence the value of σ, whereas the value of $\sigma^\star$ is mainly affected by the number of spectral points next to the BGS.

Taking into account the complete measured spectral response for the estimation of frequency uncertainties comes with several advantages such as an increased number of spectral points used for regression calculation and the consideration of spectral points less influenced by amplitude noise (within the BGS but outside the gray range in Fig. 7.2). Though, a regression function describing the entire measured spectral range is required. Literature reports on Lorentzian [3, 21, 185] and Lorentzian-Gaussian functions [3]. The BFS is one of 4 parameters describing a Lorentzian (scaling parameter, half width at half maximum, peak shift and offset) and is optimized during each least-square iteration step of the regression calculation. However, the Lorentzian function can not be explicitly solved for the BFS which would allow an error propagation calculation of BFS uncertainties as shown for parabolic functions [252]. Based upon the characteristics of amplitude noise in SBS fiber sensors, a noise model function (NMF) is designed to analytically describe the measured BGS.

Figure 7.3 summarizes the following Monte Carlo process to estimate frequency uncertainties for SBS-based fiber sensors. First, a normalized Lorentzian obtained from a measured spectral response of a SBS sensor (e.g. see Fig. 7.2) is used as reference and denoted by ref. To each spectral point of ref artificial noises are added by employing random number

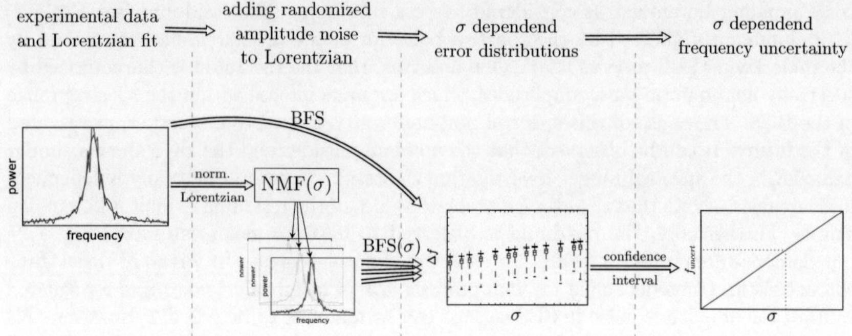

Figure 7.3: Pictographic process flow aiming to stochastically estimate the frequency uncertainty in SBS fiber sensors with σ being the standard deviation of the residuum.

generators $r(\sigma)$. Eqn. 7.2 gives a simple and straight forward NMF.

$$\text{NMF}(\sigma) = \sum_{i=1}^{N} ref_i + r_1(\sigma) + g \cdot ref_i \cdot r_2(\sigma) \qquad \text{with } g = 1/A_{res} \qquad (7.2)$$

N is the number of spectral points, σ the given standard deviation for both random number generators and r_1, r_2 the random numbers generated by the generators. r_1 is equal to the noise definition of [210] and considered as dominating for high numbers of averaging samples. Whereas $g \cdot r_2$ is used to emulate Brillouin-amplified noise amplitudes observed in the residuum for low numbers of averaging samples. g represents a theoretical Brillouin gain factor. The value of g varies with the signal-to-noise ratio along the sensing fiber and between different measurement systems. Thanks to the ergodic noise character of the residuum, g is estimated by computing a Lorentzian regression function on the absolute values of the residuum. The obtained maximum amplitude of this Lorentzian A_{res} is inversely proportional to g, since the maximum amplitude of the measurement Lorentzian is normalized to a value of 1. r_1 and r_2 as outputs of random number generators are assumed to follow a Gaussian distribution defined by σ with a mean value of zero. The obtained artificially noise-added spectrum serves as a basis for methods determining the BFS. In this section a Lorentzian regression calculation was performed. The obtained BFS was subtracted from the BFS of the reference as described in Eqn. 7.3.

$$\Delta f(\sigma) = |BFS_{ref} - BFS_{NMF}(\sigma)| \qquad (7.3)$$

In order to increase the statistical significance, 1,000 samples of Δf per given σ value were computed. On the basis of the normalized Lorentzian shown in Fig. 7.2, the Δf distribution for given σ values were computed. Figure 7.4a) shows the computational result as a boxplot for each σ value, respectively. Figure 7.4b) illustrates the distribution of frequency difference values as a histogram and confirms the Gaussian distribution of random numbers.

Figure 7.4 illustrates the direct influence of amplitude noise on the Δf distributions. The smaller the standard deviation of noise, the smaller the median value of the noise

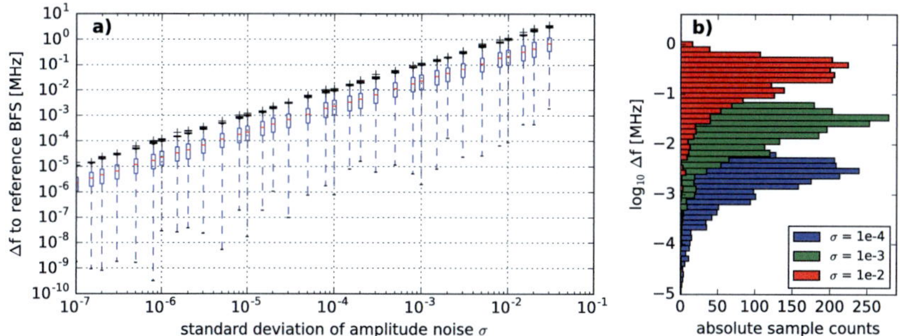

Figure 7.4: a) Distribution of frequency differences between reference spectrum and noise-added spectrum plotted against the standard deviation of the residuum. Each distribution of frequency differences is defined by a mean value (red mark), a standard deviation (blue box), minimum observed values (black mark at the bottom) and maximum observed values (black marks on top). b) Histogram of observed frequency difference values for 3 specific σ values from a) in absolute sample counts.

distributions gets. This observation is consistent with the analytic description in [210]. Furthermore, the Gaussian-shaped Δf distributions are observed to be a consequence of the Gaussian-shaped amplitude noise. The BFS frequency uncertainty f_{uncert} was defined as the maximum frequency difference within a 99.5 % confidence interval of each Δf distribution. Due to the evident log-log linear slope of f_{uncert} values against σ, a log-log linear regression function provides a simple analytic possibility to calculate the BFS frequency uncertainty for given σ values. In case of Fig. 7.2, a maximum frequency uncertainty of $f_{uncert} = 0.41$ MHz was estimated using the proposed process. Consequently, the measurement result is given as determined BFS value and f_{uncert} value, resulting in $(10,994.7 \pm 0.4)$ MHz within 99.5 % confidence interval.

Utilizing the proposed procedure to determine BFS frequency uncertainties f_{uncert}, the influence of numbers of frequency sampling points within a fixed scanning range were further investigated. The frequency sampling points of the normalized Lorentzian in Fig. 7.2 have been decreased stepwise. The computed f_{uncert} values as a function of σ for given frequency resolutions are plotted in Fig. 7.5a).

One observes that f_{uncert} is a function of σ, whereas the number of frequency sampling points has a non-linear scaling impact which is consistent to the analytic description of [210]. Figure 7.5b) emphasizes this non-linear relation by demonstrating the BFS frequency uncertainties for a given standard deviation of $\sigma = 0.005$. Taking into account both predominant impacts of f_{uncert}, the proposed calculation procedure enables a measurement optimization regarding measurement accuracy or regarding measurement time within a given frequency accuracy.

The proposed procedure to estimate the frequency uncertainty does not explicitly include changes of amplitude noises caused by setup components, noise spectral power density changes along the probe propagation path or double Rayleigh scattering generated by the probe wave. All named sources of noise appear as amplitude noise on the photo detector.

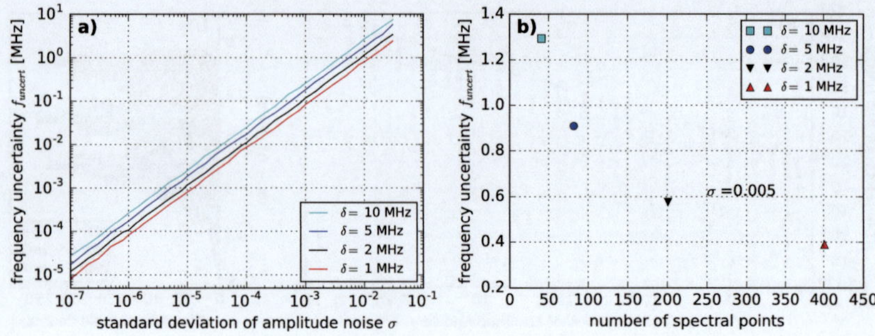

Figure 7.5:   a) Computed BFS frequency uncertainties plotted against the standard deviation of noise for given frequency resolutions within a fixed frequency range of
400 MHz. b) Computed BFS frequency uncertainties as a function of number
of spectral points for a given standard deviation of residual noise.

An analytical spectral description of σ characterizing the components and noise sources of
the optical setup can be added to the NMF to further improve the accuracy of determined
f_{uncert} values. NMFs based on covariances are conceivable, too. Compared to the analytical
state of the art approach of frequency uncertainty estimation, the presented stochastic
approach requires higher computation costs. The computing time directly scales with the
number of chosen noise samples and standard deviations. In return this estimation method
offers the possibility to quantify a frequency uncertainty for all mathematical functions
capable of describing a the entire measured spectral response of SBS-based sensors.

7.2 Frequency uncertainties in distributed measurements

The international standard for distributed fiber optic temperature measurements [239]
emphasizes the need for accurate determination of measurement uncertainties and proposes
a two step procedure for its calculation. In the first step 20 single shots of the same
measurement are averaged, followed by a spatial averaging of 51 points around the spatial
segment of interest. However, the proposed spatial averaging of [239] is only valid if
neighbored fiber segments exhibit comparable environmental influences. Therefore, the
total length of any environmental influence is assumed to be at least 50 times greater than
the spatial resolution of a distributed fiber sensor. As a consequence, spatially shorter
events tend to be underestimated in its true value and uncertainty. Since, all distributed
fiber sensors offer the opportunity to threat each spatial fiber segment as an independent
sensor, the measurement uncertainty should be determined for each individual spatial
fiber segment, too. In case of SBS-based fiber sensors the BFS uncertainty needs to be
individually obtained for each spatial fiber segment as presented in [210, 240, 241] and in
the previous section.

Different methods of BFS determination applied on the same data set lead to slightly
different BFS values. In order to compare the outcomes of BFS determination methods

to each other, the frequency uncertainty is needed to be taken into account. Based on a BOTDA raw data set the BFS value and the standard deviation of the residuum σ are determined for a Lorentzian curve fitting and a parabolic fit on a DCT output, respectively. The frequency uncertainties are estimated and compared by the analytical and stochastic approach for the parabolic fit, whereas the frequency uncertainty for the Lorentzian fitting function are presented following the stochastic approach.

A BOTDA measurement result of a ultra low-loss fiber with a length of 25 km is used to discuss distributed measurement uncertainties. The BOTDA settings were $\delta = 1$ MHz, spatial resolution of 5 m and 100 averaging samples by covering the spectral range from 10.8 GHz to 11.2 GHz. To demonstrate the capability of the novel stochastic approach another BFS determination method next to the Lorentzian regression calculation is applied on the same BOTDA raw data set. The method of choice represents the current state of the art method, in which a discrete cosine transformation (DCT) is performed on each spatial fiber segment individually before applying a parabolic regression function [253]. The number of DCT coefficients was limited to 40, which corresponds to 10 % of the number of obtained spectral points. The 40 DCT coefficients represent one mean value and 39 coefficients describing spectrally equidistant cosine functions. Limiting the number of cosine functions can be considered as a low pass filtering suppressing high frequency amplitude noise. Furthermore, the filtered BGS described by a sum of cosine functions enables the determination of the BFS using a parabolic regression function as described in [253]. The results of the DCT on a measured spectral response are shown in Fig. 7.6. One observes that the DCT-filtered BGSs show all characteristics of the original measurement result. Like Fig.7.2, the residuum of the DCT outcome demonstrates a spectral non-uniform power distribution with an ergodic character, which permits the use of the proposed stochastic approach.

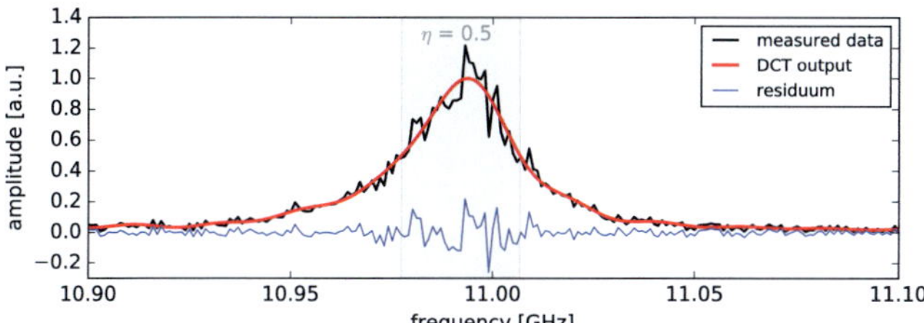

Figure 7.6: Normalized spectral response of a BOTDA fiber segment and its corresponding DCT output, which was limited to 40 coefficients. The spectral difference between both is labeled residuum.

The BFS of the BOTDA raw data set was determined by Lorentzian regression calculation and parabolic regression calculation based on a DCT output. Figure 7.7a) shows the result of the calculations. Both BFS determination methods provide slightly different results along the entire fiber length. Note, that there is a systematic difference between both BFS determination methods of approximately 0.5 - 1 MHz. Figure 7.7b) shows the spectral residui of both BFS determination methods as it serves as a basis for the stochastic approach.

In order to compare the analytic to the presented stochastic BFS uncertainty approach, both were performed on the same measurement utilizing the parabolic BFS determination

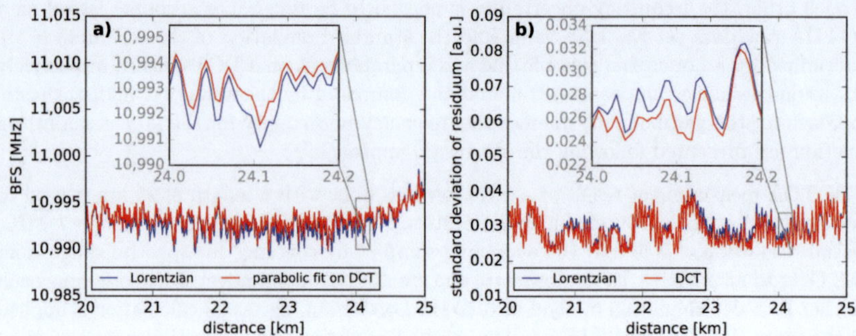

Figure 7.7: a) BFS values obtained by Lorentzian regression calculation and parabolic regression on DCT filtered measurement data. The inset provides a closer look b) Standard deviation of the residuum evaluating the whole measured spectrum range plotted against distance.

method. In case of the stochastic approach the number of computed standard deviations was set to 15 (5 values per decade in a range from $\sigma = 1\cdot10^{-1}$ to $\sigma = 1\cdot10^{-4}$) and 500 noise samples per standard deviation. The analytic approach was performed following Eqn. 7.1 and the standard deviation of amplitude noise $\sigma^\star$ was determined in the spectral range from 10.8 GHz to 10.9 GHz (101 spectral points). The results are plotted in Fig. 7.8a). Both BFS uncertainty approaches demonstrate comparable results.

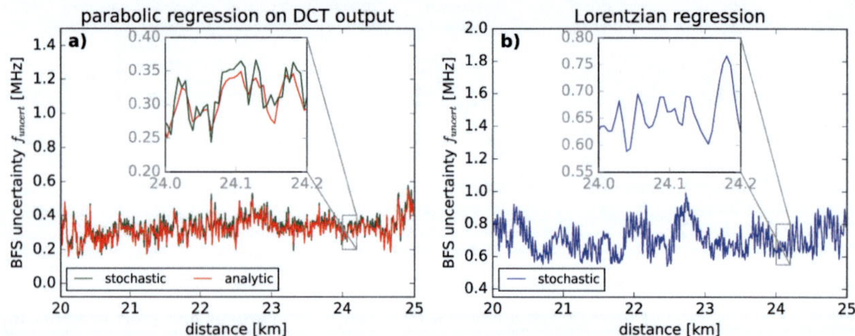

Figure 7.8: Estimated BFS uncertainties plotted against distance for both BFS determination methods, separately. The BFS uncertainties determined by the analytic approach of [210] are plotted as 3 times the estimated frequency standard deviation to obtain a comparable confidence interval as to the stochastic approach.

In the next step, the stochastic approach was applied on the same measurement raw data evaluated by Lorentzian regressions. Unfortunately, the analytic approaches of [210, 240, 241] are not valid for Lorentzian functions. Estimated BFS uncertainties for Lorentzian regression functions in a distributed measurement are shown Fig. 7.8b). The

Lorentzian BFS uncertainty determined by the stochastic approach is considered to decrease by applying noise reductions such as the presented DCT [253, 254]. However, this does not necessarily represent an improvement of measurement accuracy.

As shown for the comparison of Lorentzian and parabolic regression functions, other BFS determination methods and mathematical functions can be directly compared to each other with respect to the corresponding BFS uncertainty. In virtue of the criteria BFS uncertainty and required acquisition time, the optimum BFS determining method and hardware can be found for each specific setup based on a benchmarking system. For field applications the number of averages can be estimated to meet a given measurement uncertainty within a minimum amount of measuring time from Fig. 7.5. In addition, the minimum BFS uncertainty and therefore the limit of detection can be estimated for each SBS sensor individually by using the stochastic approach.

7.3 Figure-of-merit for BOFDA

Aiming scientists and companies to fairly evaluate the progress brought by proposed solutions improving the performance of BOTDA sensors, the FoM was introduced in 2013 [210] and is given in Eqn. 7.4. Generally, the FoM is based upon a set of experimental parameters and a proposed expression predicting the SNR in the sensor response at the receiver. Though, limitations imposed by nonlinear effects, pump depletion and parasitic fiber effects are need to be additionally considered.

$$\text{FoM} = \frac{(\alpha L_{eff})^2 \ \cdot \ exp\left\{(2 + f_l)\alpha L\right\}}{\Delta z \ \sqrt{N_{tr} N_{av}}} \frac{\sqrt{\Delta f_B \ \delta}}{f_{err}} \tag{7.4}$$

The measurement parameters are needed as well as the determined frequency error (see beginning of this chapter) and an number of averaged spatial traces. Since, the FoM was designed for BOTDA a little adjustment is needed to extent the definition for BOFDA. The time required in the time-domain to acquire one spatial trace is given by [65]:

$$t_{aq} = \frac{2 \ nL}{c} \ N_{av} \tag{7.5}$$

The equivalent to Eqn. 7.5 in frequency-domain is:

$$t_{aq} = \frac{3L}{\Delta z} \frac{1}{RBW} \tag{7.6}$$

When Eqn. 7.5 and 7.6 are set equal, follows:

$$N_{av} = \frac{c}{2n} \frac{3}{\Delta z} \frac{1}{RBW} \tag{7.7}$$

Hence, BOFDA and BOFDA measurements are comparable based on the FoM [65].

CHAPTER 8

Designing a distributed Brillouin sensor in PFGI-POF

In chapter 5 distributed Brillouin sensing in PFGI-POF is presented based on BOCDR and BOFDA. The deployment of BOTDR and BOTDA in PFGI-POF are seen critical due to high pump pulse powers. BOFDA and BOCDR benefit from the usage of modulated cw lasers, which avoid high peak powers potentially causing a PFGI-POF surface burning and fiber fuse effects [202]. However, in BOCDR the sensing range, the number of spatial segments are a function of pump power [50]. For slope-assisted BOCDR, the dynamic range is additionally affected by the pump power [177]. In case of a 14.5 m PFGI-POF the required pump power to achieve a spatial resolution of 0.96 m was reported to be 25 dBm for a slope-assisted BOCDR [175]. In order to increase the sensing range and/or spatial resolution, higher pump powers are needed, which can also lead to fiber fuse effects. The temperature sensitivity is reported to be a function of relative distance to the start of the fiber (-4.98 · 10^{-4} dB/°C/m [175]). Thus, BOFDA offers unexplored potential in POF-based sensing as the sensing range is only limited by the optical loss of the fiber under test and the lower cut-off frequency of the photo diode and/or VNA.

At the time this thesis was written, BOFDA in PFGI-POF has been only reported in [18]. The distributed measurement of a temperature-dependent BFS profile along a 20-m POF fiber sample at a nominal spatial resolution of 4 m was shown. In order to amplify the weak backscattering signal an EDFA was inserted before the photodetector. However, the EDFA did not ensure a proper operation of the setup, due to the unstable character of the backscattering signal. The origin for this power fluctuation is assumed to be the critical optical coupling of the employed FC/PC interconnections, which demonstrated a strong temperature dependency in reflective power [18]. This issue will be tackled by the usage of FC/APC interconnections (see chapter 4) and a heterodyne detection scheme to amplify the weak backscattering signal. The further enhancement potential of detection sensitivity in PFGI-POF is shown by the use optical filters demonstrated on silica multimode fibers.

As second major concern of BOFDA in PFGI-POF, the overall optical loss of the PFGI-POF was identified [18]. However, due to the choice of pump wavelength at 1550 nm in all reported distributed PFGI-POF Brillouin sensors (BOCDR and BOFDA), a significant fiber propagation loss contradicts sensing ranges above 30 m. In contrast to wavelengths around 1550 nm (130-250 dB/km [52]), wavelengths at 1319 nm demonstrate losses below 60 dB/km (see chapter 6). Therefore, longer sensing ranges at lower pump powers can be achieved, which potentially reduces the risk fiber fuse effects. However, the availability of fiber optic components around 1319 nm is limited compared to 1550 nm, which led to prototype developments of fiber optic filters in cooperation with companies and other research institutes. Thus, the comparability of measurement results to other research groups might be affected.

In this chapter a basic BOFDA setup [97, 255] (see Fig. 5.1) is adapted the meet all requirements for POF sensing. The major adaptions are discussed and their cross-influence is outlined. A setup aiming a prove of concept is presented, which has further potential to improve the obtained measurement results.

8.1 The big picture

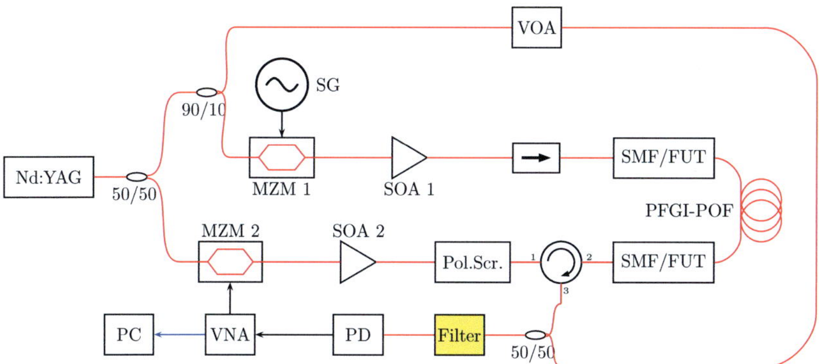

Figure 8.1: Experimental setup for BOFDA in PFGI-POF. Colors of connections between components represent different types. Red: optical fiber, black: coaxial wiring, blue: logical connection. Nd:YAG: solid state laser at 1319 nm, VOA: variable optical attenuator, MZM: Mach-Zehnder modulator, SG: signal generator, SOA: semiconductor optical amplifier, Pol.Scr.: polarization scrambler, PFGI-POF: polymer optical fiber under test, SMF/FUT: fiber coupler between PFGI-POF and single mode fiber, Filter: optional optical filter, PD: photo diode, VNA: vector network analyzer, PC: computer for data acquisition, computation and data visualization.

Figure 8.1 displays the experimental setup. The used Nd:YAG laser has a linewidth of 5 kHz at a wavelength of 1319 nm (see chapter 6). The laser output power of 22 dBm was split into two paths with a ratio of 50/50 (pump path / auxiliary path). The auxiliary path was further split in the probe path and a less power full path used as a local oscillator for the heterodyne detection. The probe path enabled the frequency shift around the Brillouin frequency of the PFGI-POF. Therefore, Mach-Zehnder modulator (MZM) 1 was modulated by an electrical signal generator. MZM 1 operated in the minimum of its transfer function to suppress the carrier (dual sideband with suppressed carrier, DSB-SC, [13, 185, 223]). Subsequently, the modulated light was amplified by a semiconductor optical amplifier (SOA). The following isolator protects the SOA and MZM from the pump path transmitting through the PFGI-POF. The pump path implemented the spatial resolution in the frequency-domain. The MZM 2 was driven by a VNA and operated in its quadrature point to ensure the linear region of the transmission function [13, 185]. The SOA boosted the frequency modulated signal, proving a power regulation independent of the probe path. A polarization scrambler in the pump path ensures an interaction between the counter-propagating pump and probe waves in the PFGI-POF. Therefore, insensitive fiber parts due to orthogonal pump and probe waves were avoided [256].

The choice of lasing wavelength at 1319 nm affected the selection of components. The only commercially available fiber optic amplifiers at 1319 nm with sufficiently high output powers are based on semiconductors, which have greater noise figures and gain ripples compared to an EDFA at 1550 nm. Therefore, higher amplitude noises are expected compared to a BOFDA setup at 1550 nm. In addition, external modulators capable of driving modulation frequencies in the lower GHz-range at 1319 nm are limited to MZMs. State of the art distributed Brillouin setup for silica fibers at 1550 nm employ acousto-optic modulators and MZMs with increased extinction ratios (ERs), in order to rise the signal-to-noise ratio in the obtained BGSs [11]. Both employed MZMs in the final setup demonstrate an ER above 27 dB (see Fig. 8.2). Figure 8.2 was further used to determine the required BIAS voltages and RF output powers as well as the degrees of modulation (see chapter 5), respectively.

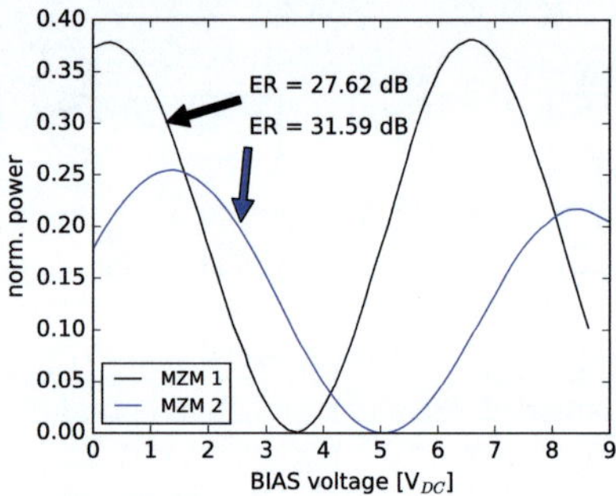

Figure 8.2:   Normalized MZM output power with respect to the applied BIAS voltage. MZM1 and MZM2 refer to the modulators shown in Fig. 8.1.

As described in [18, 21, 111] and chapter 3, the maximum SBS power level is generated by injecting the laser light into the fundamental mode of the PFGI-POF. In order to realize the required accurate coupling, handcrafted FC/APC connectors (see chapter 4) were used at both ends of the PFGI-POF in the setup. The 8° angled cut prevents light reflections from the interface from traveling back into the fiber. Taking into consideration the extreme high coherence length of the Nd:YAG laser, FC/APC connectors were employed in the entire setup to reduce power fluctuations caused by interface reflections, which can act as standing wave interferometers.

By employing a self-heterodyne detection the sensitivity of a BOTDA sensor was increased and experimentally demonstrated a 10.75-dB enhancement in the SNR compared to direct-detection systems [143]. In addition, distributed measurements with self-heterodyne detection demonstrated a decreased BFS standard deviation [143], indicating an enhanced accuracy of BFS estimation. Neglecting the Rayleigh backscattering, the interaction of the amplified probe wave and a local oscillator electrical fields can be generally expressed by:

$$\mathbf{E}_d(t, f) = \mathbf{E}_S g_B(f_S, z)\, exp\left\{j\left[2\pi f_S t + \Psi_B(f_S, z)\right]\right\} + \mathbf{E}_{LO}\, exp\left\{j2\pi f_{LO} t\right\} \tag{8.1}$$

where $\mathbf{E}_S$ and $\mathbf{E}_{LO}$ are the complex field amplitudes of the probe wave and local oscillator, respectively. g_B and Ψ_B are the Brillouin gain and Brillouin phase shift. Consequently, the detection current at the photodetector can be written as:

$$i(t) = R_{ph}P_d = 2R_{ph}[1 + g_B(f_S, z)]\sqrt{P_S P_{LO}}\, cos\left[2\pi f_{IF} t + \Psi_{IF} - \Psi_B(f_S, z)\right] \tag{8.2}$$

where R_{ph} is the responsibility of the photodetector as well as P_S and P_{LO} are the optical power of the probe wave and local oscillator, respectively. The phase and frequency difference between probe wave and local oscillator are given as Ψ_{IF} and f_{IF}. The detection current scales with the power of the local oscillator. Thus, an all optical and tunable signal amplification is achieved. The self-heterodyne detection is an essential part of a BOFDA sensor in PFGI-POF. In Fig. 8.1 a VOA in the local oscillator path enables to control the signal amplification.

The self-heterodyne detection affects all optical powers propagating in direction of the probe wave. Hence, Eqn. 8.1 and Eqn. 8.2 need to consider further signals to represent the actual laboratory setup such as the upper side band and the carrier of the probe wave, the Rayleigh backscattering of the pump wave including its sidebands and frequency noise induced by spontaneous emissions of the SOAs. The power of the local oscillator is spread to all named optical signals according to their power levels. In order to focus on the relevant spectral band, a narrow bandwidth fiber optical filter is added to setup (see Fig. 8.1). In addition, the filter helps to reduce relative intensity noise of the photodetector as it attenuates spectral signals with high powers such as the local oscillator and the Rayleigh backscattering of the pump path. A closer view on the influence of fiber optical filters is given in section 8.3.

All optical signals detected at the PD are measured and collected at the VNA to obtain a S21 trace (see chapter 5). The PC was used for data acquisition, control of the measurement components, calculation of the fiber transfer functions and evaluation the obtained spectral responses for each spatial fiber segment individually.

8.2 Fiber transfer function

The crucial step in BOFDA is the determination of the fiber transfer function. However, errors to the obtained magnitude or phase for each f_m value are a result of (i) random noise in the electrical signal, (ii) random noise in the optical signal, (iii) nonlinearities of deployed setup components and (iv) within the Brillouin interaction along the sensing fiber [74]. In order to outline the influence of a spectral noise $\eta(jf)$ on the transfer function

$G(jf))$, the inverse Fourier transfer of a noisy signal is shown. For the sake of simplicity, f_D will not appear. The degraded pulse response $g_\eta(t)$ is reconstructed as follows:

$$g_\eta(t) = \int_{-\infty}^{+\infty} [G(jf) + \eta(jf)] \ e^{jft} df \tag{8.3}$$

First, assuming an additive error in the magnitude at frequency f_0 with correct phase information Ψ leads to the measured value: :

$$G(jf_0) = [G(jf_0) + A_\eta] \ e^{j\Psi} \tag{8.4}$$

As a consequence of the error in magnitude, the obtained inverse Fourier transform changes to [74]:

$$g_\eta(t) = g(t) + \frac{A_\eta}{\pi} cos(2\pi f_0 t) \tag{8.5}$$

Thus, an error in the magnitude measurement (e.g. due to polarization scrambling of the pump wave) results in an oscillation with the frequency f_0 in the time domain, which affects the spatially resolved spectral responses. This oscillation of measurement values can be reduced by temporal averaging.

Assuming the measured transfer function at the frequency f_0 to be affected in magnitude and phase, Eqn. 8.4 changes to:

$$G(jf_0) = [G(jf_0) + A_\eta] \ e^{j\Psi + \Psi_\eta}. \tag{8.6}$$

Accordingly, the measured value of $G_\eta(jf_0)$ is need to be expressed by [74]:

$$\underbrace{G_\eta(jf_0)}_{\text{measured value}} = \underbrace{\frac{1}{2cos(\Psi_\eta)}}_{\text{error term}} \left[\underbrace{|G(jf_0)e^{j\Psi}}_{\text{actual value}} + \underbrace{A_\eta e^{j\Psi}}_{\text{error term}} \right] + \underbrace{\frac{|G(jf_0)| + A_\eta}{2cos(\Psi_\eta)} e^{j[\Psi + 2\Psi_\eta]}}_{\text{error term}} \tag{8.7}$$

The degraded pulse response is the inverse Fourier transform of Eqn. 8.7 [74]:

$$\begin{aligned} g_\eta(t) = \ & g(t) + \frac{|G(jf_0|) + A_\eta}{2\pi cos(\Psi_\eta)} [cos(2\pi f_0 t + \Psi) + cos(2\pi f_0 + \Psi + 2\Psi_\eta)] \\ & - \frac{|G(jf_0)|}{\pi} cos(2\pi f_0 t + \Psi) \end{aligned} \tag{8.8}$$

Again, the obtained temporal response is a superposed oscillation with the frequency f_0, weighted by the magnitude value $|G(jf_0)| + A_\eta$. Both spectral noise assumptions resulted in a degradation of obtained temporal responses. Equation 8.8 underlines that phase inaccuracies have a more severe impact compared to a defect in magnitude measurement. The occurrence of such errors are dependent on the configuration of the recording system for $G(jf)$ and where the reference signal for phase detection is drawn from [74]. A self-heterodyne detection scheme as presented in Fig. 8.1 is considered more robust to fluctuations compared to an external electrical oscillator.

To minimize the influence of each setup component on the obtained results, a calibration of the setup was performed [74]. Without any calibration the measured transfer function $G'(jf)$ is assumed the be a product of the desired transfer function of the sensing fiber with all frequency responses of components in the signal path:

$$G'(jf) = G_{PD}(jf) \cdot G_{MZM2}(jf) \cdot G_{VNA}(jf) \cdot \underbrace{G_{fiber}(jf)}_{desired} \cdot \ldots \qquad (8.9)$$

By short-circuiting the fiber under test all frequency responses apart of the fiber under test were measured in a calibration function. Thus, the measurement of a BOFDA trace could be realized without a second photodetector. Only the small additional expense of a single calibration measurement is needed followed by a division of magnitudes and a subtraction of phase angles from the measured transfer function. The impacts of calibration can be summed up in three points: (i) The group delay of all setup components involved are set to zero, which enables the fiber evaluation starting at 0 m. (ii) Nonlinear frequency responses of setup components are linearized, which leads to more accurate results. (iii) The Brillouin gain for spatial segments exceeding the fiber under test are set to roughly zero.

Even though a calibration minimizes the impact of setup components, a nonlinear distortion of high modulation frequencies f_m is need to considered. As described in chapter 5, an ideal pulse needed for the spatial resolution in BOFDA systems represents an sinc-shaped spectral equivalent. In order to suppress side slopes of the sinc function, a Kaiser window (with a chosen β-value of 5) is applied before performing the inverse Fourier transform. However, this Kaiser window affects high values of f_m especially for short fibers and when setting fine spatial resolutions (see Eqn. 5.12). Future BOFDA setups could tackle this issue by designing a software-based notch filter matching the desired f_m range, individually.

VNAs are capable of measuring the real and imaginary part of the transfer function, simultaneously. Therefore, evaluations regarding the Brillouin gain amplitude and the Brillouin phase shift can be achieved. Furthermore, the complex Brillouin gain spectrum [250] could be spatially resolved. In order to reduce complexity and costs of the setup hardware in later commercial devices, the transfer function's real part can be measured only. As a consequence, only the Brillouin gain amplitude is detected. For the sake of completeness, reference [250] shows, that the nonlinear Brillouin phase shift affects the Brillouin gain amplitude, too.

To obtain the pulse response without the imaginary part of the transfer function, the real part of the complex transfer function is expressed by:

$$\Re\left\{G(jf)\right\} = \frac{G(jf) + G^*(jf)}{2}. \qquad (8.10)$$

Equation 8.10 is valid for low-pass filtered signals satisfying the condition $f_m < RBW < 4f_m$ [74]. This condition essentially helps with the choice of the VNA's resolution bandwidth. Assuming real signals for $g(t)$ and $g(t) = g(-t)$, the transformation pairs are $G(jf) \bullet\!\!-\!\!\circ g(t)$ and $G^*(jf) \bullet\!\!-\!\!\circ g^*(-t)$. Thus, the superposition of the obtained pulse response $g(t)$ with its mirrored image $g(-t)$ result in:

$$g_{Re}(t) = \frac{g(t) + g(-t)}{2}. \qquad (8.11)$$

For the sake of simplicity and in order to reduce the phase noise influence (see Eqn. 8.8), all presented measurement results in this thesis were obtained following Eqn. 8.11. According to [74] the frequency step width is required to be half as long as in a full complex measurement when implementing Eqn. 8.11.

An alternative approach to determining the fiber transfer function via an inverse Fourier transfrom is presented in [257]. The proposed reconstruction technique is based on a numerical model of possible BFS distribution profiles from which a hypothetical transfer function is chosen. The process of extracting the most suited transfer function is based on means of a numerical solution for simplifications of the three differential equations (see Eqn. 2.9). The chosen transfer function is then iteratively altered in its parameters to achieve similarity to the measured transfer function. This alteration yields to the physical distribution of Brillouin gain along the fiber. However, the algorithm of this approach is considered computationally expensive as in each iteration, the set of differential equation has to be solved, and convergence is not assured for all arbitrary BFS distributions [74].

8.3 Optical filtering

A variety of fiber optical filters can be used to select transmission bands. Exemplarily named are gas cambers providing molecule absorption bands, thin film filters and optical resonators, such as fiber Bragg gratings (FBGs) and Fabry-Perot-interferometers (FPIs). The Brillouin frequency shift of PFGI-POF is around 3 GHz and roughly in the range of 11.5 GHz to 13.5 GHz for silica multimode fibers at 1550 nm and 1319 nm pump wavelength, respectively. Hence, extremely narrow-bandwidth filters are required to separate the pump wave and the probe wave from each other. Optical resonators can be engineered to meet these requirements with sufficiently high filter depths and low insertion loss for specific wavelengths. In the 1550 nm wavelength range ultra narrow-bandwidth FBGs and FPIs are commercially available and are employed in numerous distributed Brillouin fiber sensors [11, 13, 185]. In cooperation with external partners FPIs and FBGs at the wavelength of 1319 nm were engineered to prototype level and their impact on BOFDA measurement results is demonstrated.

Fiber Bragg grating

Fiber Bragg gratings are based on Fresnel reflection and interference. Portions of light are reflected at regions with different refractive indexes. In FBGs the refractive index alternates over a defined length. When light travels through a fiber with a periodically modulated refractive index, a spectrally narrow-band with beam is reflected at the wavelength λ_B, if the difference in the propagation constants of the reflected is equal to the spatial frequency of the grating (see Braggs law in chapter 2). The spatial period of the alternation Λ and the effective refractive index of the grating n_e define the reflected Bragg wavelength λ_B [117, 258].

$$\lambda_B = 2 n_e \Lambda \tag{8.12}$$

The linewidth and filter depth of a FBG strongly depend on the FBG length, the distribution of refractive index changes along the grating and the spectral target shape [258]. In cooperation with Fraunhofer Heinrich-Hertz-Institute Goslar six narrow-bandwidth FBGs

Table 8.1: Summary of manufactured FBGs. Each FBG prototype demonstrated an insertion loss of roughly 3.5 dB at their center wavelength.

Grating ID	Type	center wavelength	3 dB linewidth	Filter depth
1	Gaussian	1318.8 nm	150 pm / 25.8 GHz	11.8 dB
2	Gaussian	1319.0 nm	150 pm / 25.8 GHz	12.5 dB
3	Gaussian	1319.0 nm	125 pm / 21.5 GHz	13.7 dB
4	π-shifted	1319.24 nm	161 pm / 27.5 GHz	9.2 dB
5	π-shifted	1319.2 nm	30 pm / 5.1 GHz	8.5 dB
6	π-shifted	1319.1 nm	80 pm / 14.2 GHz	7.9 dB

were designed and manufactured (see Tab. 8.1). Due to the small amount of FBGs, the fabrication scheme utilizing phase masks was unfeasible. Instead, each scattering center was written individually by an ultra-violet femto-second laser. Two different FBG types were investigated aiming to realize two filter schemes. For FBG type named Gaussian in Tab. 8.1, the scattering centers were inscribed equidistant to each other and the refractive modulation of scattering centers followed a 6th order Gaussian function along the inscription length. Therefore, the FBGs demonstrate higher filter depths compared to all π-shifted FBGs. However, this comes at the costs of broader 3 dB linewidths. In addition to the Gaussian FBGs, pi-shifted FBGs were manufactured. π-shifted are generally following the spectral response of Gaussian FBGs but with an additional destructive interference exactly at the center wavelength. However, the filter depth of the π-shifted FBGs is limited, due to the prototype character of the FBGs.

The target center wavelength of all manufactured FBGs was designed to be slightly smaller than the lasing of the Nd:YAG (1319.33 nm). All center wavelengths in Tab. 8.1 are given for unstrained fibers and at 22 °C. The center wavelength of a FBG depends on temperature changes ($\partial\lambda_B/\partial T \approx 10$ pm/K in SMF [117]) and tensile strain ($\partial\lambda_B/\partial S \approx 1$ pm m/μm in SMF [117]) [117, 258]. Thus, the FBGs were spectrally aligned with the Brillouin frequency of the fiber under by applying tensile strain to the FBG. For this purpose a 3-axis stage with a manual operation of the micrometer misalignment was used.

FBG ID 5 demonstrated the narrowest 3 dB linewidth of all manufactured FBGs. However, the temperature and strain stabilization of this π-shifted FBGs was practically impossible over the required time span of a BOFDA measurement. Thus, FBG ID 3 was used for the BOFDA measurements as it demonstrated the narrowest linewidth and highest filter depth of all Gaussian FBGs. However, the filtering bandwidth drastically increased compared to the π-shifted FBG preventing a spectral separation of the setups pump and probe wave.

In order to investigate the influence of FBG ID 3, a 8.8-km silica GI-MMF (see Tab. 4.1) without any temperature oder strain influences was deployed as fiber under test in the BOFDA setup as shown in Fig. 8.1. The measurement parameters were set to pump power of 7 dBm, probe power of -5 dBm, heterodyne path power -5 dBm, 500 averaging samples, 3 kHz resolution bandwidth, 4 MHz spectral resolution and 27 m spatial resolution resulted in a measurement time of 28 minutes. The results are shown in Fig. 8.3 and the comparison of estimated BFS and normalized probe power are displayed in Fig. 8.4.

In Fig. 8.3 it can be observed, that the FBG reduces the detected optical power without influencing the shape of the spectral responses. This optical attenuation can be attributed to the insertion loss of the FBG, the decreased power outside the FBG transition band

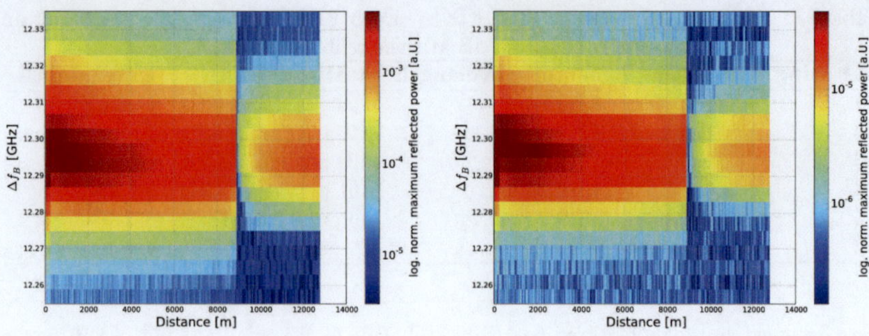

Figure 8.3: BOFDA measurement results of a 8.8-km silica GI-MMF without (left) and with (right) the employment of the FBG ID 3.

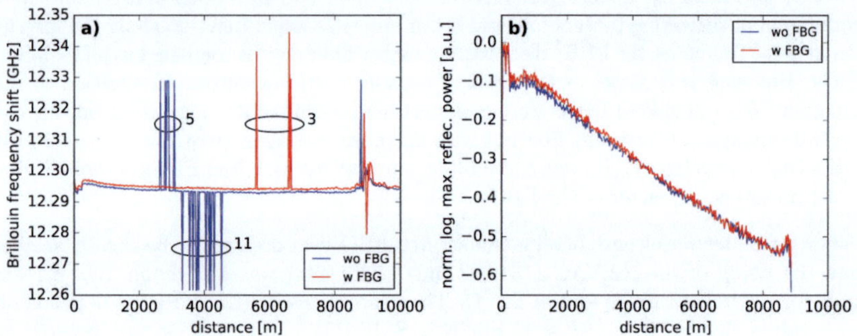

Figure 8.4: Comparison of estimated Lorentzian fit parameters extracted from Fig. 8.3. a) Brillouin frequency against distance. b) Normalized logarithmic probe power plotted as a function of distance.

and the reduced relative intensity noise of the photodetector caused by the SOAs and pump power. Figure 8.4 underlines the benefit of a FBG employment. The number of fiber segments with incorrect BFS estimations is reduced by a factor of 5 without affecting the normalized detected power in each spatial fiber segment. Thus, the impact of the SOAs on measurement results should not be neglected.

Fabry-Perot interferometer

The FPI is an optical cavity, which allows light passing through when they are in resonance. Resonant wavelengths of a FPI λ_{FPI} are expressed by:

$$\lambda_{FPI} = \frac{2\Lambda n}{u} \tag{8.13}$$

where Λ is the geometrical distance between both reflecting surfaces, n the refractive index and $u \in \mathbb{N}$. The FPI can be considered a multi-beam-interferometer with negligible attenuation. The incident light intensity I_0 is split into two parts - a reflected intensity $I_{reflected}$ and a transmitted intensity $I_{transmitted}$. The ratio of splitting is predominantly defined by the power reflection coefficient $R_p = I_{reflected}/I_0$. The FPI is characterized by a regular phase difference Ψ_{FPI} between two consecutive beams.

$$\Psi_{FPI} = 2\pi n \Lambda / \lambda + \Delta\Psi \tag{8.14}$$

$\Delta\Psi$ denotes phase jumps due to reflection. Thanks to the neglected attenuation the reflected and transmitted light intensities are expressed by the Airy functions [259]:

$$I_{reflected} = I_0 \frac{U \, sin^2(\Psi_{FPI}/2)}{1 + U \, sin^2(\Psi_{FPI}/2)} \tag{8.15}$$

$$I_{transmitted} = I_0 \frac{1}{1 + U \, sin^2(\Psi_{FPI}/2)} \tag{8.16}$$

where $U = 4R_p/(1-R_p)^2$. According to Eqn. 8.13 and the Airy functions, the FPI provides multiple equidistant optical modes. The spectral spacing between two successive intensity maxima / minima is called free spectral range (FSR) and is expressed by:

$$FSR_{FPI} = \frac{c}{2n\lambda} \tag{8.17}$$

The full-width at half-maximum in a transmission spectrum is defined as given in Eqn. 8.18. The ratio of FSR and maxima linewidth at full-width at half-maximum is called finesse.

$$FWHM_{FPI} = \frac{2FSR}{\pi U} \tag{8.18}$$

In cooperation with Micron Optics, Inc. a piezo-driven tunable FPI was designed and manufactured to improve the filtering outcome compared to FBGs. The final FPI is characterized by a center wavelength of the first optical mode at 1316.4 nm in absence of any BIAS voltage, a FSR of 3 GHz, a finesse of 583, a bandwidth of 5.16 GHz, a maximum optical input power of 10 dBm and an insertion loss of 2.4 dB, 2.2 dB and 2.2 dB for the first 3 resonant modes at 1310 nm, respectively. In contrast to the prototype FBGs, the FPI offered a filter depth greater than 40 dB and an easy voltage-controlled handling of the wavelength tuning. FPIs with increased finesse would have easily led to narrower bandwidths at the expense of reduced maximum optical input powers. In order to ensure the function of the FPI under unpredictable power fluctuations a higher maximum input power was chosen over a narrow bandwidth. However, temperature changes of the FPI prototype affected the center wavelength, too. Thus, the FPI was kept in a polystyrene box to minimize temperature-induced wavelength disturbances.

The influence of the applied BIAS voltage on the first optical mode of the FPI was investigated in order to accurately control the filter in setup. For this purpose a SOA and the Nd:YAG laser are coupled into a VOA with the help of an 50/50 power coupler. The VOA added a 50 dB attenuation to the combined signal before entering the OSA. Figure 8.5 displays the obtained results for selected BIAS voltages.

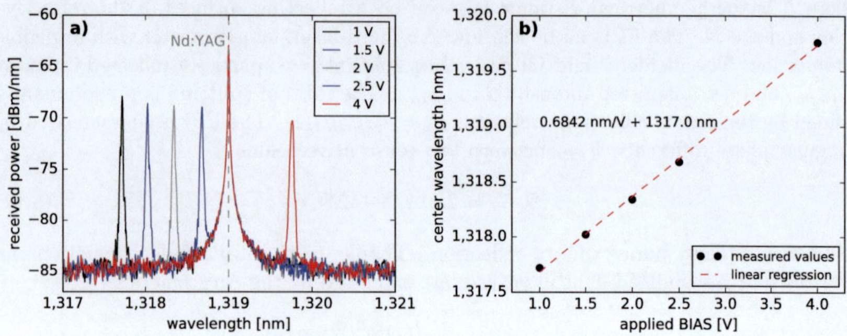

Figure 8.5: a) Transmission spectra of FPI with applied BIAS voltage. b) FPI center wavelength of the first mode as a function of applied BIAS voltage including a linear regression function computed for applied BIAS voltages from a).

As expected according to Eqn. 8.13, the center wavelength of the FPI is a linear function of the applied BIAS voltage. Taking into account the slope of the linear regression (0.6842 nm/V) and the voltage accuracy of the power supply (1 mV), a maximum wavelength uncertainty of 0.6842 pm ($\approx$ 118 MHz around 1319 nm) results.

Aiming the BOFDA setup to operate with PFGI-POF much shorter sensing fiber lengths as shown in Fig.8.3 are necessary, due to the high fiber propagation loss. In general, the reduction of sensing fiber length is seen critical as the much higher optical powers are required to generate stimulated Brillouin scattering. However, the SOA need high operating currents, which lead to an increased probability of spontaneous photon emissions. In addition, it needs to be pointed out, that the gain of SOA decreases with increasing output powers. Consequently, the maximum output power of a SOA also depends on its input power. Increased optical powers in the fiber under test also increase the Rayleigh backscattering power of the pump as well as the expected input power of the FPI and photodetector. The latter causes relative intensity noise which again negatively affects the signal-to-noise ratio of the obtained spectral responses.

The employment the precise FPI prototype in the BOFDA setup (see Fig. 8.1) enabled the reduction of the GI-MMF length down to 200 m even without heterodyne detection. The measurement parameters are set to a pump power of 21 dBm, probe power of 8dBm, 200 averaging samples, 3 kHz resolution bandwidth, 4 MHz spectral resolution and 0.21 m spatial resolution. The total measurement lasted 36 min. The results are shown in Fig. 8.6.

The achieved results underline the significant influence of optical filters in the setup. At comparable measurement times and compared to the results obtained with FBGs, the number of averaging samples was reduced by a factor of 5 and the number of spatial points was increased by a factor of 5 by employing a FPI. On the other hand, the standard deviation of BFS noise along the fiber under test does increase. This observation can be associated to higher pump powers necessary to generate SBS in short fibers and its resulting impacts on the obtained spectral responses. Consequently, the BFS uncertainty increased.

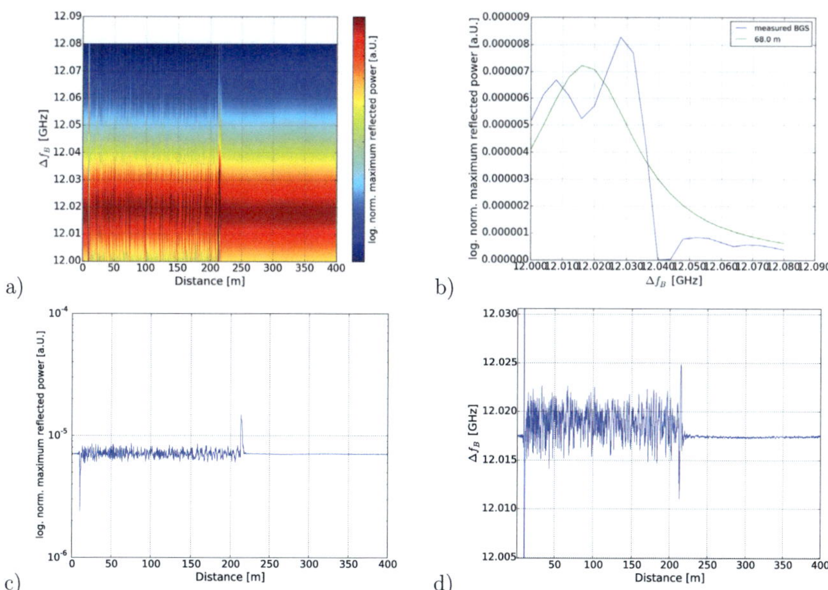

Figure 8.6: BOFDA evaluation of a 200-m silica GI-MMF with 50 μm core diameter. a) Computed backscattering power as a function of distance and frequency detuning. b) Measured spectral response of an evaluated spatial fiber segement and its corresponding Lorentzian regression function. c) Computed Lorentzian peak value of each spatial fiber segment plotted as a function of distance. d) Computed BFS profile along the fiber under test.

Furthermore, the spatial fiber segments beyond the fiber under test are observed to demonstrate the mean values of the spectral responses along the FUT. The begin and end of the FUT can be clearly identified considering the BFS profile and power profile. The SMFs connecting the GI-MMF, are characterized to have a non-overlapping BGS with the FUT. Therefore, the this behavior is seen to originate from the spatially low-pass character of BOFDA. Additional high-pass filters in the data evaluation could suppress this behavior as reported in [13, 184].

8.4 Experimental BOFDA results in POF

A PFGI-POF is employed as a fiber under test in the setup shown in Fig. 8.1. Based on the achieved results in a 200-m silica GI-MMF, the 200-m PFGI-POF, which was used for fiber characterization (see chapter 6), is installed. However, any BOFDA attempt failed due to the high insertion loss of the PFGI-POF. The total loss along the 200-m PFGI-POF at 1310 nm is measured to be 16.3 dB, which is consistent with [21, 22] and chapters 4 and 6. Therefore, the fiber length is reduced to 86 m, which results in a decreased total loss along the PFGI-POF of 11.9 dB. Compared to the reported BOFDA system in PFGI-POF [18], a significant reduction of fiber insertion loss is achieved. The reported insertion losses of

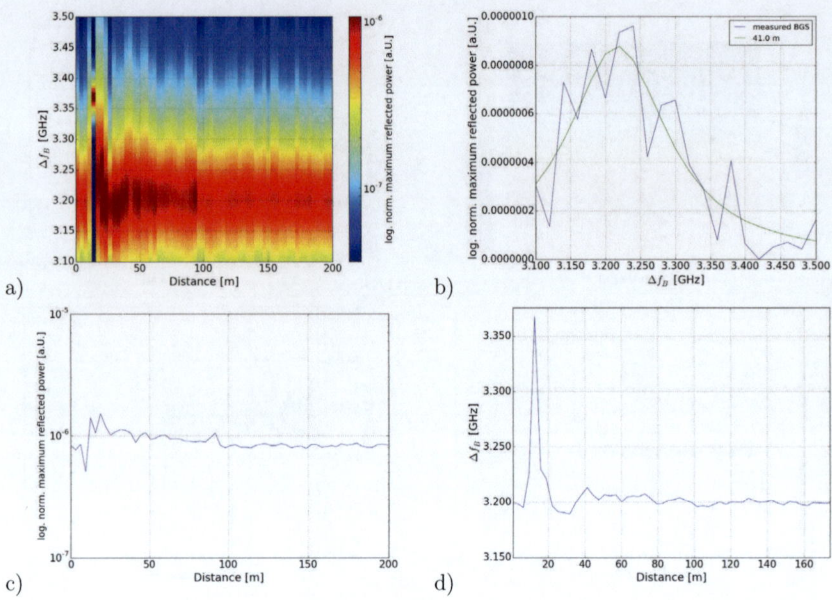

Figure 8.7: BOFDA evaluation of a 86-m PFGI-POF with a self-heterodyne detection scheme only. a) Computed backscattering power as a function of distance and frequency detuning. b) Measured spectral response of an evaluated spatial fiber segement and its corresponding Lorentzian regression function. c) Computed Lorentzian peak value of each spatial fiber segment plotted as a function of distance. d) Computed BFS profile along the fiber under test.

the SMF-POF and POF-SMF interfaces at a wavelength of 1550 nm are 8 dB and 18.5 dB, respectively [18].

The powers in the pump and probe path of the setup are stepwise increased to find the optimum operation point of the BOFDA system. The pump and probe power are empirically determined to be 21 dBm and 14 dBm, respectively. In order to demonstrate the significant influence of the FPI on the measurement results, a comparison of two BOFDA measurements are aimed - one without the FPI and the second with the FPI. For this purpose the PFGI-POF was coiled on a fiber spool under slight tensile strain apart of a 5 m long fiber section which remained strain-free. The entire fiber length was assumed to have a uniform temperature distribution along the fiber of 22 °C. The results for the self-heterodyne detection without the FPI are shown in Fig. 8.7. The measurement parameters of 1000 averaging samples, 3 kHz resolution bandwidth, 20 MHz spectral resolution and 3.4 m spatial resolution resulted in a measurement time of 32 minutes.

It can be observed, that the proposed BOFDA setup is capable to spatially resolve Brillouin gain spectra along a 86-m long PFGI-POF even without the usage of a FPI. The Brillouin linewidths around 120 MHz of the slightly strain fiber section (Fig. 8.7a) and b)) are consistent with the results presented in chapter 6. The reflection of the pump wave at the fiber end can be found in Fig. 8.7c). Though, smaller spatial resolutions might enable a more precise localization of the fiber end. Furthermore, the strain-free fiber section was successfully detected at 16.75 m. The obtained spectral response of the strain-free fiber segment demonstrates a Lorentzian shape with an extremely narrow linewidth and a BFS shifted towards higher frequencies compared to the rest of the fiber. The BFS was measured to be 3.357 GHz (see Fig. 8.7d)), which would correspond to a temperature of 12 °C in a non-strained PFGI-POF following Eqn. 6.1. However, the temperature discrepancy between measured and actual values as well as the extremely narrow bandwidth of the obtained spectral response indicate this segment to be not reliable. Repeated measurements demonstrate similar results. As the strain-free fiber section is roughly the size of the set spatial resolution and due to the limited signal to noise ratio of the spectral response, the BOFDA algorithm is expected to provide inaccurate results. Though, the spatial localization of the event remains correct.

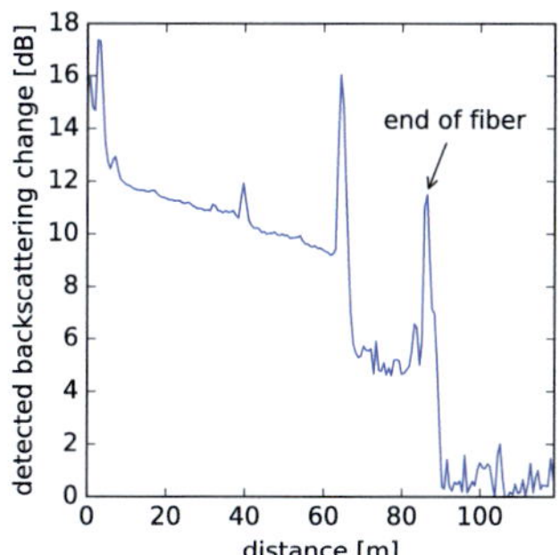

Figure 8.8: OTDR backscatter trace at a wavelength of 1311 nm with 1 m spatial resolution.

Unfortunately, the comparing measurement with the FPI employed can not be presented as the insertion loss of the PFGI-POF drastically increased after the first measurement. The PFGI-POF facets remained intact and without any visual damage when investigated with a high resolution digital microscope. The origin of the increased insertion loss is found by the help of an OTDR. Figure 8.8 shows the obtained OTDR trace of the 86-m PFGI-POF after the BOFDA measurement. It can be observed, that a strong light reflection occurs around 64 m and the fiber end reflection can be located at 86 m. The light reflection at 64 m causes an decreased backscattering power after its location until the fiber end. With the help of a red fiber laser illuminating the PFGI-POF at both fiber ends simultaneously, the strong light reflection was observed to be occurring along a very short distance of less than 1 mm. Thus, the major reflection peak in Fig. 8.8 is assumed to be a scattering center inscribed by the BOFDA setup during the measurement. Similar inscription processes lead to the fiber fuse effect when the higher optical powers are applied [202]. However, the exact localization along the fiber of such inscribed scattering centers can not be predicted at the moment [202].

The exposure time of a PFGI-POF in the BOFDA setup and its influence on inscribed scattering canters was further investigated. A PFGI-POF from a second manufacturer (Fontex from ASAHI Glass [1]) was investigated next to the GigaPOF-50SR from Chromis Fiber Optics used for all measurements so far. Both fibers have got the same core diameter

---

[1] http://www.lucina.jp/eg_fontex/

of 50 μm and a comparable graded-index profile. To provide similar test conditions, both fibers are annealed for 72 hours at 70 °C, 95 % relative humidity and gradually cooled to room temperature within 24 hours in the same climate cabinet, simultaneously. OTDR measurements are performed after each exposure time in the BOFDA setup. The results for the Fontex fibers are shown in Fig. 8.9 with the lead-in SMF patch cord marked red.

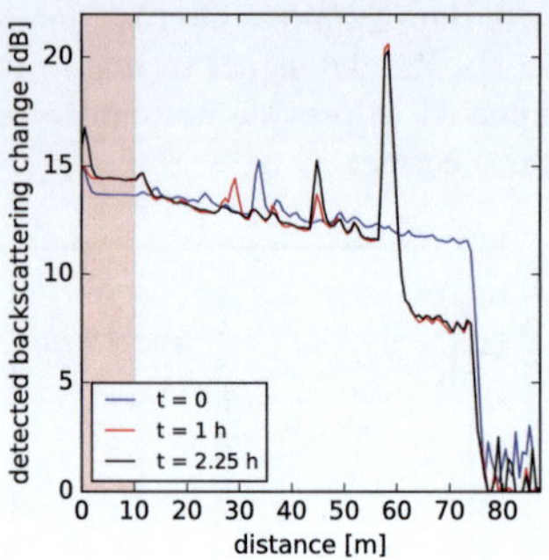

Figure 8.9: OTDR traces of one PFGI-POF after given exposure times.

As for the GigaPOF-50SR, scattering centers are inscribed into the Fontex fiber. The first exposure time is observed to predominantly inscribe the major scattering center (see backscattering reflection peak at 57 m when comparing the traces 't = 0' and 't = 1 h'), as additional exposure times do not significantly contribute to its reflectivity (see backscattering reflection peak at 57 m when comparing the traces 't = 1 h' and 't = 2.25 h'). Comparable result are achieved for the GigaPOF-50SR fiber. This behavior is assumed to result from the two powerful light sources interfering constructively, which would cause a local intensity exceeding the damaging threshold of the fiber. The first inscribed scattering center separates the pump and probe wave of the BOFDA setup by increasing transmission loss for both light paths. Therefore, the probability of constructive interfering light path, which can exceed the threshold of fiber damaging, is minimized. Any further inscribed scattering centers are a result of interference with backscattered light from the first dominant scattering center. Though, the reflectivity of these secondary scattering centers is observed to be weaker compared to the first inscribed scattering centers (see backscattering reflection peak at 45 m in all three traces). The extremely high coherence length of the employed Nd:YAG laser is considered to significantly contribute to local optical intensities damaging the PFGI-POF. The influence of the laser coherence length in BOFDA systems with respect to the inscription of scattering centers in PFGI-POF should be further investigated.

Assuming a comparable dynamic range to the commercial OTDR system, a BOFDA trace at a fixed frequency difference between pump and probe path should provide a similar result as shown in Fig. 8.8. Therefore, any scattering center affect the pump and probe power distribution along the PFGI-POF and reduce measurement accuracy. In addition, scattering centers can be found in the PFGI-POF itself (see Fig. 6.14a)). These intrinsic scattering centers are associated to the non-uniform distributions of dopants in the graded-index profile, a crystallization of the CYTOP core material, impurities of the polymers and the manufacturing process of the fibers (see chapter 3). Thus, the scattering centers are randomly distributed along the fiber. In addition, the intrinsic scattering centers are random in their sizes and density which correlates to the reflected intensities. Intrinsic scattering centers exceeding 7 dB peaks in OTDR traces are observed to be rather common than rare. Next to the influence on the pump and probe wave power distribution along the fiber, intrinsic scattering centers are need to be considered as blind spots for distributed Brillouin sensors. The light scattering at the intrinsic centers is independent of the frequency difference between pump and probe wave. Therefore, no target function of any regression calculation can estimate a BFS for spectral responses corresponding to spatial fiber segments including a dominant scattering center.

However, the inscription of scattering centers into the PFGI-POF prevented a comparison measurement with a FPI when employing the same fiber. In addition, the confirmation of the temperature and strain coefficients (see chapter 6) in the designed BOFDA setup is not possible due to the same issue. An improved manufacturing process providing less intrinsic scattering centers and lower fiber loss would help to reduce the Brillouin threshold and therefore the required optical powers in the setup. On the other hand, the reduction of measurement time is essential. Shorter exposure times lead to less dominant first inscribed scattering centers and might enable the PFGI-POF to be measured more than once. Shorter BOFDA measurement times can be achieved by replacing the VNA with an all-digital acquisition of the complex transfer function. Alternatively, bigger fiber core diameters (commercially available up to 120 μm) might reduce the danger of inscribed scattering center as the fundamental mode diameter also increases and therefore to minimize the impact of density and dopant concentration fluctuations along the fiber. Though, higher Brillouin thresholds are expected with greater core diameters. Thus, further investigations based on the results in this thesis are necessary to implement a fully functional BOFDA setup in PFGI-POF.

CHAPTER 9

Summary and Outlook

The aim of this thesis was to design and implement a distributed Brillouin sensor based on BOFDA for perfluorinated polymer optical fibers. Compared to silica fibers, polymer optical fibers offer advantages such as significantly higher break down strain, fracture toughness and humidity sensitivity. Prior to this work, distributed Brillouin sensors in polymer optical fibers were reported to be predominantly use the correlation-domain to determine locations along the fiber. The correlation-domain approach developed during the duration of this PhD project and successfully implemented dynamic Brillouin sensing techniques in PFGI-POF. The only presented BOFDA in PFGI-POF achieved a 4 m nominal spatial resolution in a 20 m PFGI-POF and was capable of measuring a temperature change of 15 K. Irrespective of the domain approach, all reported distributed Brillouin sensors utilize laser sources at a wavelength around 1550 nm, which corresponds to a fiber attenuation of 150 - 250 dB/km. Consequently, high optical pump powers are required to generate stimulated Brillouin scattering on PFGI-POF. In order to potentially reduce the required optical powers and extend reported sensing ranges, an operation wavelength of 1319 nm was chosen as it corresponds to significantly smaller fiber losses of less than 60 dB/km.

A more in depth understanding of the physical background of stimulated Brillouin scattering - especially in graded-index multimode fibers - form the basis for a later adaptation of the BOFDA scheme (see chapters 2 and 3). The light propagation in PFGI-POF differs from commonly used silica fibers as mode coupling effects are need to be considered. Therefore, theoretical assumptions are made in order to describe the SBS process in PFGI-POFs (see chapter 3). The dominant excitation of the fundamental mode is key to generate SBS in PFGI-POFs. Combined with the modal selection for data evaluation, a fiber sensor based on SBS can be designed.

Unfortunately, the coupling efficiency between a given pair of modes and fibers is not measurable when using a multimode fiber. Therefore, the analytic description of coupling coefficients enable the setting of boundary conditions for the hand-crafted SMF - PFGI-POF interconnections. The built interconnections ensure reproducible measurement results at given excited and evaluated optical modes with a significantly reduced insertion losses (see chapter 4).

The literature survey presented in chapter 5 provides a basic overview on distributed SBS sensing. The Brillouin optical frequency-domain analysis and its theoretical background as well as SBS sensors in PFGI-POF are discussed in detail.

The entire relationships between the SBS parameters (backscattering power, Brillouin linewidth and BFS) and the environmental parameters strain, temperature and humidity are thoroughly presented, respectively (see chapter 6). A considerable sensitivity to humidity is proven and its origin is discussed. All the presented results are relevant for temperature and humidity fiber sensors in PFGI-POF such as fiber Bragg gratings and distributed sensors based on Rayleigh or stimulated Brillouin scattering.

Any BFS uncertainties result in uncertainties of the measured environmental parameters. To estimate BFS uncertainties an analytical model based on a parabolic fit is commonly used as state of the art. In order to expand the variety of underlaying mathematical regression functions, a stochastic approach is presented (see chapter 7). However, further investigations regarding the spectral noise distribution within the measured Brillouin gain spectra are necessary to substantiate the proposed approach.

Finally, the implementation of the BOFDA setup for PFGI-POF is introduced. Special emphasis is placed on the understanding of error propagation through the inverse Fourier transform to obtain the spatially distributed power profile. The improvement of BOFDA sensitivity by employing an optical filter is outlined. **A distributed Brillouin fiber sensor in PFGI-POF is demonstrated achieving a sensing range of 86 m with 3.4 m spatial resolution**. Though, the full potential of the proposed setup with equipped optical filters remains unclear due to scattering centers inscribed during the BOFDA measurements. In addition, intrinsic scattering centers in the fiber core are considered to limit the further practical realization of long range SBS sensors in PFGI-POF (see chapter 8). On the other hand, the setup demonstrates spatial resolutions as fine as 0.21 m in a 200-m long graded-index silica fiber without the usage of the heterodyne detection scheme.

Outlook

A great effort was made to advance the research on distributed Brillouin sensors in PFGI-POF. In order to further pursuit for efficient sensor implementation and later practical application requires to address the following issues, if the research is to be continued:

- Replacing the VNA with a digital frequency synthesizer could further increase the dynamic range and shorten the measurement times as demonstrated for incoherent OFDR [260].

- The combination of FPI and self-heterodyne detection probably will further decrease the detection noise level and therefore increase the sensor range and accuracy.

- The increase of the fiber core diameter raises the Brillouin threshold, but could significantly decrease the light intensity, which causes inscribed scattering centers.

- Changing the probe path modulation scheme to phase modulation [261] or carrier-suppressed single-side band generation (SSB SC) [255, 262] could potentially reduce the overall optical path power. Therefore, the probability of inscribing scattering centers during the BOFDA measurement is decreased.

- A further evaluation of the obtained VNA raw data sets could provide further outcomes regarding the non-linear Brillouin phase in PFGI-POF. The obtained result could help to further understand the influence of mode coupling. Furthermore, the spatially resolved evaluation of the complex Brillouin gain spectrum in PFGI-POF might enable a reduction of BFS uncertainty through theoretical analysis, simulations and experiments as shown for silica SMFs in [250].

- The integration of an additional 1x2 50/50 coupler combining pump and probe path before entering the fiber under test would transform the setup to an L-BOFDA [188]. The maximum sensing length is expected to be reduced compared to classical BOFDA, but it would provide the opportunity for a single end fiber access operation. Taking into account the coupling losses between PFGI-POF and SMF, a reduction of interface numbers is considered beneficial. However, as L-BOFDA utilizes the Fresnel reflection at the far end of the fiber, the fiber fuse effect is expected to occur starting at certain PFGI-POF lengths. In addition, the concept of L-BOFDA have not been proven in laboratory setups or any field tests, yet.

Next to the Brillouin coefficients, chapter 6 provides a thorough investigation of the influence of humidity changes on optical properties of the PFGI-POF. Various humidity-related effects in PFGI-POFs have been reported in the last years including measured attenuation and length changes as well as Brillouin frequency and Bragg wavelength shifts. The four aforementioned methods could serve as a basis for a distributed and/or quasi-distributed humidity sensing and are compared in Tab. 9.1 [263].

Table 9.1: Comparison of (quasi-)distributed fiber sensor techniques and effects for humidity sensing in PFGI-POF. [263]

	Distributed attenuation change	Distributed length change	Quasi-distributed FBG wavelength shift	Distributed Brillouin frequency shift
Method / complexity	Difference between two computed slopes from Rayleigh backscatter trace / moderate	Correlation-based length change of signal patterns from Rayleigh backscatter trace [34,264]/ moderate	Interrogator [128, 265] / moderate	complex [18, 49, 51]
Sensitivity to humidity	Spectral dependency and function of temperature [22, 94]	CHE is independent of temperature $7.4\,10^{-6}$ %r.h. [22] and no function of wavelength [22]	14.7 pm/%r.h. at 30 °C [128]. Influence of temperature needs to be investigated.	Analytical description [21] and theoretically given as - 108 kHz/%r.h. [22]

Table 9.1 – *Continued from previous page*

	Distributed attenuation change	Distributed length change	Quasi-distributed FBG wavelength shift	Distributed Brillouin frequency shift
Spatial resolution	Depending on spatial resolution of measurement system (e.g. 5 cm with Luna 4413 [22] and 3.7 cm with incoherent OFDR [34])		Quasi distributed [128, 265]	Depending on spatial resolution of measurement system(e.g. 10 cm, BOCDR [51])
Sensing length	Based on wavelength-dependent transmission loss (e.g. 500 m for 1311 nm [266]) and backscatter measurement method		Limited by fiber propagation loss, number of FBGs and insertion loss of each FBG	Based on wavelength-dependent transmission loss (e.g. 20 m at 1550 nm [51])
Real-time capability	Measurement time from seconds [34] up to minutes [22]	Measurement time from seconds [34] up to minutes [22]	Measurement time from microseconds up to minutes	Measurement time in range of seconds [51]
	Sensor response time is a function of fiber core diameter, fiber cladding material, fiber cladding layer thickness, fiber jacket material and value of humidity change [94, 267, 268]. Response times are reported in the range of minutes to days [94, 267, 268]			
Cross-sensitivity to temperature and possibility for separation	Humidity sensitivity is a function of temperature for humidity sensitive wavelengths [22, 94].	No distinction to temperature (CTE $= 22{,}7$ 10^{-6} K^{-1}) [22].	Distinction between strain and temperature was presented in [93].	No distinction to temperature [21].
Cross-sensitivity to strain and possibility for separation	No distinction to strain [94].	No distinction to strain [22].	No distinction to strain [22].	No distinction to strain [22].

Cited from [263]:
"As shown in table 9.1, the distinction of humidity between temperature and strain is not natively given for all four presented techniques and effects. Possible approaches to enable humidity measurements are plausibly seen in the combination of measurement

methods. Furthermore, an additional humidity-insensitive sensing fiber for strain and/or temperature might provide required data to extract humidity information along a PFGI-POF. In case of attenuation change and Brillouin frequency shift, the employment of a second sensing wavelength could allow a wavelength-depended evaluation as shown for PMMA [23]. However, all presented solution approaches for humidity sensing require data evaluation of multiple sensors and sensor systems. Multi-variable models and neuronal networks are seen as promising evaluation methods to provide a distributed humidity sensor in PFGI-POF. [...] Thus, the field of application specifies the most suitable distributed fiber sensor with respect to required humidity sensitivity/accuracy, requested sensing fiber length, expected temperature and strain ranges, requirements regarding measurement time versus real time capability, fiber installation conditions and possibility of second sensing method or fiber."

Bibliography

[1] G. P. Agrawal, *Nonlinear Fiber Optics.* Academic press, 2007.

[2] R. W. Boyd, *Nonlinear Optics.* Elsevier, 2003.

[3] T. Schneider, *Nonlinear Optics in Telecommunications.* Springer Science & Business Media, 2013.

[4] S. Chin, L. Thévenaz, J. Sancho, S. Sales, J. Capmany, P. Berger, J. Bourderionnet, and D. Dolfi, "Broadband true time delay for microwave signal processing, using slow light based on stimulated Brillouin scattering in optical fibers," *Optics Express*, vol. 18, no. 21, pp. 22 599–22 613, 2010.

[5] S. Preußler, N. Wenzel, R.-P. Braun, N. Owschimikow, C. Vogel, A. Deninger, A. Zadok, U. Woggon, and T. Schneider, "Generation of ultra-narrow, stable and tunable millimeter-and terahertz-waves with very low phase noise," *Optics Express*, vol. 21, no. 20, pp. 23 950–23 962, 2013.

[6] I. Carr and D. Hanna, "Performance of a Nd: YAG oscillator/ampflifier with phase-conjugation via stimulated Brillouin scattering," *Applied Physics B: Lasers and Optics*, vol. 36, no. 2, pp. 83–92, 1985.

[7] I. S. Grudinin, A. B. Matsko, and L. Maleki, "Brillouin lasing with a CaF2 whispering gallery mode resonator," *Physical Review Letters*, vol. 102, no. 4, p. 043902, 2009.

[8] H. H. Diamandi and A. Zadok, "Ultra-narrowband integrated Brillouin laser," *Nature Photonics*, vol. 13, no. 1, p. 9, 2019.

[9] A. Zadok, A. Eyal, and M. Tur, "Stimulated Brillouin scattering slow light in optical fibers," *Applied Optics*, vol. 50, no. 25, pp. E38–E49, 2011.

[10] D. Garcus, T. Gogolla, K. Krebber, and F. Schliep, "Brillouin optical-fiber frequency-domain analysis for distributed temperature and strain measurements," *Journal of Lightwave Technology*, vol. 15, no. 4, pp. 654–662, 1997.

[11] A. Motil, A. Bergman, and M. Tur, "State of the art of Brillouin fiber-optic distributed sensing," *Optics & Laser Technology*, vol. 78, pp. 81–103, 2016.

[12] A. Loayssa, J. Urricelqui, H. Iribas, and J. José, *Dynamic to Long-Range Measurements in Sensors for Diagnostics and Monitoring.* CRC Press, 2018.

[13] T. Kapa, A. Schreier, and K. Krebber, "A 100 km BOFDA assisted by first-order bi-directional Raman amplification," *Sensors*, vol. 19, no. 7, p. 1527, 2019.

[14] M. A. Soto, X. Angulo-Vinuesa, S. Martin-Lopez, S.-H. Chin, J. D. Ania-Castanon, P. Corredera, E. Rochat, M. Gonzalez-Herraez, and L. Thévenaz, "Extending the real remoteness of long-range Brillouin optical time-domain fiber analyzers," *Journal of Lightwave Technology*, vol. 32, no. 1, pp. 152–162, 2014.

[15] R. Bernini, G. Persichetti, E. Catalano, L. Zeni, and A. Minardo, "Refractive index sensing by Brillouin scattering in side-polished optical fibers," *Optics Letters*, vol. 43, no. 10, pp. 2280–2283, 2018.

[16] R. Cohen, Y. London, Y. Antman, and A. Zadok, "Brillouin optical correlation domain analysis with 4 millimeter resolution based on amplified spontaneous emission," *Optics Express*, vol. 22, no. 10, pp. 12 070–12 078, 2014.

[17] A. Zarifi, B. Stiller, M. Merklein, and B. Eggleton, "High resolution Brillouin sensing of micro-scale structures," *Applied Sciences*, vol. 8, no. 12, p. 2572, 2018.

[18] A. Minardo, R. Bernini, and L. Zeni, "Distributed temperature sensing in polymer optical fiber by BOFDA," *IEEE Photonics Technology Letters*, vol. 26, no. 4, pp. 387–390, 2014.

[19] P. J. Thomas and J. O. Hellevang, "A fully distributed fibre optic sensor for relative humidity measurements," *Sensors and Actuators B: Chemical*, vol. 247, pp. 284–289, 2017.

[20] P. Thomas and J. Hellevang, "A high response polyimide fiber optic sensor for distributed humidity measurements," *Sensors and Actuators B: Chemical*, vol. 270, pp. 417–423, 2018.

[21] A. Schreier, A. Wosniok, S. Liehr, and K. Krebber, "Humidity-induced Brillouin frequency shift in perfluorinated polymer optical fibers," *Optics Express*, vol. 26, no. 17, pp. 22 307–22 314, 2018.

[22] A. Schreier, S. Liehr, A. Wosniok, and K. Krebber, "Investigation on the influence of humidity on stimulated Brillouin backscattering in perfluorinated polymer optical fibers," *Sensors*, vol. 18, no. 11, 2018.

[23] S. Liehr, M. Breithaupt, and K. Krebber, "Distributed humidity sensing in PMMA optical fibers at 500 nm and 650 nm wavelengths," *Sensors*, vol. 17, no. 4, p. 738, 2017.

[24] P. Stajanca, L. Mihai, D. Sporea, D. Neguţ, H. Sturm, M. Schukar, and K. Krebber, "Effects of gamma radiation on perfluorinated polymer optical fibers," *Optical Materials*, vol. 58, pp. 226–233, 2016.

[25] P. Stajanca and K. Krebber, "Radiation-induced attenuation of perfluorinated polymer optical fibers for radiation monitoring," *Sensors*, vol. 17, no. 9, p. 1959, 2017.

[26] S. Liehr, S. Münzenberger, and K. Krebber, "Wavelength-scanning coherent otdr for dynamic high strain resolution sensing," *Optics Express*, vol. 26, no. 8, pp. 10 573–10 588, 2018.

[27] M.-T. Hussels, S. Chruscicki, A. Habib, and K. Krebber, "Distributed acoustic fibre optic sensors for condition monitoring of pipelines," in *Sixth European Workshop on Optical Fibre Sensors*, vol. 9916. International Society for Optics and Photonics, 2016, p. 99162Y.

[28] K. Hicke and K. Krebber, "Towards efficient real-time submarine power cable monitoring using distributed fibre optic acoustic sensors," in *Optical Fiber Sensors Conference (OFS), 2017 25th*. IEEE, 2017, pp. 1–4.

[29] P. Stajanca, S. Chruscicki, T. Homann, S. Seifert, D. Schmidt, and A. Habib, "Detection of leak-induced pipeline vibrations using fiber—optic distributed acoustic sensing," *Sensors*, vol. 18, no. 9, p. 2841, 2018.

[30] P. Rohwetter, R. Eisermann, and K. Krebber, "Distributed acoustic sensing: Towards partial discharge monitoring," in *24th International Conference on Optical Fibre Sensors*, vol. 9634. International Society for Optics and Photonics, 2015, p. 96341C.

[31] G. Bashan, H. H. Diamandi, Y. London, E. Preter, and A. Zadok, "Optomechanical time-domain reflectometry," *Nature Communications*, vol. 9, no. 1, p. 2991, 2018.

[32] T. M. Daley, B. M. Freifeld, J. Ajo-Franklin, S. Dou, R. Pevzner, V. Shulakova, S. Kashikar, D. E. Miller, J. Goetz, J. Henninges *et al.*, "Field testing of fiber-optic distributed acoustic sensing (DAS) for subsurface seismic monitoring," *The Leading Edge*, vol. 32, no. 6, pp. 699–706, 2013.

[33] Y. Antman, A. Clain, Y. London, and A. Zadok, "Optomechanical sensing of liquids outside standard fibers using forward stimulated Brillouin scattering," *Optica*, vol. 3, no. 5, pp. 510–516, 2016.

[34] S. Liehr, M. Wendt, and K. Krebber, "Distributed strain measurement in perfluorinated polymer optical fibres using optical frequency domain reflectometry," *Measurement Science and Technology*, vol. 21, no. 9, p. 094023, 2010.

[35] Y. Mizuno and K. Nakamura, "Potential of Brillouin scattering in polymer optical fiber for strain insensitive high accuracy temperature sensing," *Optics Letters*, vol. 35, no. 23, pp. 3985–3987, 2010.

[36] J. Zubia and J. Arrue, "Plastic optical fibers: An introduction to their technological processes and applications," *Optical Fiber Technology*, vol. 7, no. 2, pp. 101–140, 2001.

[37] A. Tagaya, Y. Koike, T. Kinoshita, E. Nihei, T. Yamamoto, and K. Sasaki, "Polymer optical fiber amplifier," *Applied Physics Letters*, vol. 63, no. 7, pp. 883–884, 1993.

[38] S. O'Keeffe, A. F. Fernandez, C. Fitzpatrick, B. Brichard, and E. Lewis, "Real-time gamma dosimetry using PMMA optical fibres for applications in the sterilization industry," *Measurement Science and Technology*, vol. 18, no. 10, p. 3171, 2007.

[39] B. M. Quandt, L. J. Scherer, L. F. Boesel, M. Wolf, G.-L. Bona, and R. M. Rossi, "Body-monitoring and health supervision by means of optical fiber-based sensing systems in medical textiles," *Advanced Healthcare Materials*, vol. 4, no. 3, pp. 330–355, 2015.

[40] C. Leitão, V. Ribau, V. Afreixo, P. Antunes, P. André, J. L. Pinto, P. Boutouyrie, S. Laurent, and J. M. Bastos, "Clinical evaluation of an optical fiber-based probe for the assessment of central arterial pulse waves," *Hypertension Research*, vol. 41, no. 11, p. 904, 2018.

[41] A. G. Leal-Junior, C. R. Díaz, C. Leitão, M. J. Pontes, C. Marques, and A. Frizera, "Polymer optical fiber-based sensor for simultaneous measurement of breath and heart rate under dynamic movements," *Optics & Laser Technology*, vol. 109, pp. 429–436, 2019.

[42] G. Wandermur, D. Rodrigues, R. Allil, V. Queiroz, R. Peixoto, M. Werneck, and M. Miguel, "Plastic optical fiber-based biosensor platform for rapid cell detection," *Biosensors and Bioelectronics*, vol. 54, pp. 661–666, 2014.

[43] R. N. Lopes, D. M. Rodrigues, R. C. Allil, and M. M. Werneck, "Plastic optical fiber immunosensor for fast detection of sulfate-reducing bacteria," *Measurement*, vol. 125, pp. 377–385, 2018.

[44] R. C. Allil, A. Manchego, A. Allil, I. Rodrigues, A. Werneck, G. C. Diaz, F. T. Dino, Y. Reyes, and M. Werneck, "Solar tracker development based on a POF bundle and Fresnel lens applied to environment illumination and microalgae cultivation," *Solar Energy*, vol. 174, pp. 648–659, 2018.

[45] Y. Mizuno and K. Nakamura, "Experimental study of Brillouin scattering in perfluorinated polymer optical fiber at telecommunication wavelength," *Applied Physics Letters*, vol. 97, no. 2, p. 1103, 2010.

[46] Y. Mizuno, P. Lenke, K. Krebber, and K. Nakamura, "Characterization of Brillouin gain spectra in polymer optical fibers fabricated by different manufacturers at 1.32 and 1.55 um," *IEEE Photonics Technology Letters*, vol. 24, no. 17, pp. 1496–1498, 2012.

[47] Y. Mizuno, M. Kishi, K. Hotate, T. Ishigure, and K. Nakamura, "Observation of stimulated Brillouin scattering in polymer optical fiber with pump–probe technique," *Optics Letters*, vol. 36, no. 12, pp. 2378–2380, 2011.

[48] Y. Mizuno, W. Zou, Z. He, and K. Hotate, "Proposal of Brillouin optical correlation-domain reflectometry (BOCDR)," *Optics Express*, vol. 16, no. 16, pp. 12 148–12 153, 2008.

[49] Y. Mizuno, N. Hayashi, H. Fukuda, K. Y. Song, and K. Nakamura, "Ultrahigh-speed distributed Brillouin reflectometry," *Light: Science & Applications*, vol. 5, no. 12, p. e16184, 2016.

[50] Y. Mizuno, H. Lee, and K. Nakamura, "Recent advances in Brillouin optical correlation-domain reflectometry," *Applied Sciences*, vol. 8, no. 10, p. 1845, 2018.

[51] H. Lee, N. Hayashi, Y. Mizuno, and K. Nakamura, "Slope-assisted Brillouin optical correlation-domain reflectometry using polymer optical fibers with high propagation loss," *Journal of Lightwave Technology*, vol. 35, no. 11, pp. 2306–2310, Jun 2017.

[52] Y. Koike, *Fundamentals of Plastic Optical Fibers.* John Wiley & Sons, 2014.

[53] L. Brillouin, "Diffusion de la lumière et des rayons X par un corps transparent homogène - Influence de l'agitation thermique," *Annales de Physique*, vol. 9, no. 17, pp. 88–122, 1922.

[54] E. Gross, "Change of wave-length of light due to elastic heat waves at scattering in liquids," *Nature*, vol. 126, no. 3171, p. 201, 1930.

[55] R. Chiao, C. Townes, and B. Stoicheff, "Stimulated Brillouin scattering and coherent generation of intense hypersonic waves," *Physical Review Letters*, vol. 12, no. 21, p. 592, 1964.

[56] P. Kaiser and H. Astle, "Low-loss single-material fibers made from pure fused silica," *Bell System Technical Journal*, vol. 53, no. 6, pp. 1021–1039, 1974.

[57] S. E. Miller, E. A. Marcatili, and T. Li, "Research toward optical-fiber transmission systems," *Proceedings of the IEEE*, vol. 61, no. 12, pp. 1703–1704, 1973.

[58] E. Ippen and R. Stolen, "Stimulated Brillouin scattering in optical fibers," *Applied Physics Letters*, vol. 21, no. 11, pp. 539–541, 1972.

[59] R. G. Smith, "Optical power handling capacity of low loss optical fibers as determined by stimulated Raman and Brillouin scattering," *Applied Optics*, vol. 11, no. 11, pp. 2489–2494, 1972.

[60] T. Sonehara and H. Tanaka, "Forced Brillouin spectroscopy using frequency-tunable continuous wave lasers," *Physical Review Letters*, vol. 75, no. 23, p. 4234, 1995.

[61] D. Pinnow, S. Candau, J. LaMacchia, and T. Litovitz, "Brillouin scattering: viscoelastic measurements in liquids," *The Journal of the Acoustical Society of America*, vol. 43, no. 1, pp. 131–142, 1968.

[62] G. Scarcelli and S. H. Yun, "Confocal Brillouin microscopy for three-dimensional mechanical imaging," *Nature Photonics*, vol. 2, no. 1, p. 39, 2008.

[63] S. Speziale, H. Marquardt, and T. S. Duffy, "Brillouin scattering and its application in geosciences," *Reviews in Mineralogy and Geochemistry*, vol. 78, no. 1, pp. 543–603, 2014.

[64] H. Ohno, H. Naruse, M. Kihara, and A. Shimada, "Industrial applications of the BOTDR optical fiber strain sensor," *Optical Fiber Technology*, vol. 7, no. 1, pp. 45–64, 2001.

[65] A. Minardo, R. Bernini, R. Ruiz-Lombera, J. Mirapeix, J. M. Lopez-Higuera, and L. Zeni, "Proposal of Brillouin optical frequency domain reflectometry (BOFDR)," *Optics Express*, vol. 24, no. 26, pp. 29 994–30 001, 2016.

[66] S. Preußler, *Bandwidth Reduction of Stimulated Brillouin Scattering and Applications in Optical Communication.* Cuvillier Verlag, 2016.

[67] R. Engelbrecht, *Nichtlineare Faseroptik: Grundlagen und Anwendungsbeispiele.* Springer-Verlag, 2015.

[68] M. Damzen, V. Vlad, A. Mocofanescu, and V. Babin, *Stimulated Brillouin Scattering: Fundamentals and Applications.* CRC press, 2003.

[69] W. Gardner, "Appendix on nonlinearities for G. 650," *ITU Document COM*, pp. 15–273, 1996.

[70] D. Cotter, "Observation of stimulated Brillouin scattering in low-loss silica fibre at 1.3 um," *Electronics Letters*, vol. 18, no. 12, pp. 495–496, 1982.

[71] P. Bayvel and P. Radmore, "Solutions of the SBS equations in single mode optical fibres and implications for fibre transmission systems," *Electronics Letters*, vol. 26, no. 7, pp. 434–436, 1990.

[72] M. O. Van Deventer and A. J. Boot, "Polarization properties of stimulated Brillouin scattering in single-mode fibers," *Journal of Lightwave Technology*, vol. 12, no. 4, pp. 585–590, 1994.

[73] L. Thevenaz, *Limitations Caused by Nonlinear Effects.* Cost 241 - Final Report, 1998.

[74] N. Nöther, A. Wosniok, K. Krebber, and E. Thiele, "A distributed fiber optic sensor system for dike monitoring using brillouin frequency domain analysis," in *Proc. SPIE*, vol. 7003, 2008, pp. 700 303–1.

[75] FF-GI-B075 – Graded-Index Plastic Optical Fiber. [Online]. Available: https://store.fiberfin.com/media/custom/upload/File-1403209861.pdf

[76] Y. Koike and M. Asai, "The future of plastic optical fiber," *NPG Asia Materials*, vol. 1, no. 1, p. 22, 2009.

[77] E. Nihei, T. Ishigure, N. Tanio, and Y. KOIKE, "Present prospect of graded-index plastic optical fiber in telecommunication," *IEICE Transactions on Electronics*, vol. 80, no. 1, pp. 117–122, 1997.

[78] W. R. White, L. L. Blyler, R. Ratnagiri, and M. Park, "Manufacture of perfluorinated plastic optical fibers," in *Optical Fiber Communication Conference.* Optical Society of America, 2004, p. ThI1.

[79] Y. Koike and T. Ishigure, "High-bandwidth plastic optical fiber for fiber to the display," *Journal of Lightwave Technology*, vol. 24, no. 12, pp. 4541–4553, 2006.

[80] C. Lethien, C. Loyez, J.-P. Vilcot, N. Rolland, and P. A. Rolland, "Exploit the bandwidth capacities of the perfluorinated graded index polymer optical fiber for multi-services distribution," *Polymers*, vol. 3, no. 3, pp. 1006–1028, 2011.

[81] A. Doğan, B. Bek, N. Cevik, and A. Usanmaz, "The effect of preparation conditions of acrylic denture base materials on the level of residual monomer, mechanical properties and water absorption," *Journal of Dentistry*, vol. 23, no. 5, pp. 313–318, 1995.

[82] Graded-index polymer optical fiber (GI-POF). [Online]. Available: https://www.thorlabs.com/catalogPages/1100.pdf

[83] Corning® ClearCurve® multimode optical fiber - product information. [Online]. Available: https://www.corning.com/media/worldwide/coc/documents/Fiber/PI1468_07-14_English.pdf

[84] T. Ishigure, Y. Koike, and J. W. Fleming, "Optimum index profile of the perfluorinated polymer-based GI polymer optical fiber and its dispersion properties," *Journal of Lightwave Technology*, vol. 18, no. 2, pp. 178–184, 2000.

[85] P. J. Decker, A. Polley, J. H. Kim, and S. E. Ralph, "Statistical study of graded-index perfluorinated plastic optical fiber," *Journal of Lightwave Technology*, vol. 29, no. 3, pp. 305–315, 2010.

[86] S. Kawakami and H. Tanji, "Evolution of power distribution in graded-index fibres," *Electronics Letters*, vol. 19, no. 3, pp. 100–102, 1983.

[87] R. Shi, C. Koeppen, G. Jiang, J. Wang, and A. Garito, "Origin of high bandwidth performance of graded-index plastic optical fibers," *Applied Physics Letters*, vol. 71, no. 25, pp. 3625–3627, 1997.

[88] A. Garito, J. Wang, and R. Gao, "Effects of random perturbations in plastic optical fibers," *Science*, vol. 281, no. 5379, pp. 962–967, 1998.

[89] S. E. Golowich, W. White, W. A. Reed, and E. Knudsen, "Quantitative estimates of mode coupling and differential modal attenuation in perfluorinated graded-index plastic optical fiber," *Journal of Lightwave Technology*, vol. 21, no. 1, p. 111, 2003.

[90] A. Polley and S. E. Ralph, "Mode coupling in plastic optical fiber enables 40-Gb/s performance," *IEEE Photonics Technology Letters*, vol. 19, no. 16, pp. 1254–1256, 2007.

[91] W. White, M. Dueser, W. Reed, and T. Onishi, "Intermodal dispersion and mode coupling in perfluorinated graded-index plastic optical fiber," *IEEE Photonics Technology Letters*, vol. 11, no. 8, pp. 997–999, 1999.

[92] P. Stajanca, O. Cetinkaya, M. Schukar, P. Mergo, D. J. Webb, and K. Krebber, "Molecular alignment relaxation in polymer optical fibers for sensing applications," *Optical Fiber Technology*, vol. 28, pp. 11–17, 2016.

[93] A. G. Leal-Junior, A. Theodosiou, C. Marques, M. J. Pontes, K. Kalli, and A. Frizera, "Compensation method for temperature cross-sensitivity in transverse force applications with FBG sensors in POFs," *Journal of Lightwave Technology*, vol. 36, no. 17, pp. 3660–3665, Sep 2018.

[94] S. Liehr, *Fibre Optic Sensing Techniques Based on Incoherent Optical Frequency Domain Reflectometry*. BAM-Dissertationsreihe, 2015.

[95] H. Yang, S. J. Lee, E. Tangdiongga, C. Okonkwo, H. P. van den Boom, F. Breyer, S. Randel, and A. Koonen, "47.4 Gb/s transmission over 100 m graded-index plastic optical fiber based on rate-adaptive discrete multitone modulation," *Journal of Lightwave Technology*, vol. 28, no. 4, pp. 352–359, 2010.

[96] A. Kobyakov, S. Kumar, D. Q. Chowdhury, A. B. Ruffin, M. Sauer, S. R. Bickham, and R. Mishra, "Design concept for optical fibers with enhanced SBS threshold," *Optics Express*, vol. 13, no. 14, pp. 5338–5346, 2005.

[97] A. Wosniok, "Untersuchungen zur Unterscheidung der Einflussgrößen Temperatur und Dehnung bei Anwendung der verteilten Brillouin-Sensorik in der Bauwerksüberwachung," Ph.D. dissertation, Technische Universität Berlin, Fakultät II - Mathematik und Naturwissenschaften, 2013.

[98] W.-W. Ke, X.-J. Wang, and X. Tang, "Stimulated Brillouin scattering model in multi-mode fiber lasers," *IEEE Journal of Selected Topics in Quantum Electronics*, vol. 20, no. 5, pp. 305–314, 2014.

[99] B. Ward and M. Mermelstein, "Modeling of inter-modal Brillouin gain in higher-order-mode fibers," *Optics express*, vol. 18, no. 3, pp. 1952–1958, 2010.

[100] A. Li, Q. Hu, and W. Shieh, "Characterization of stimulated Brillouin scattering in a circular-core two-mode fiber using optical time-domain analysis," *Optics Express*, vol. 21, no. 26, pp. 31 894–31 906, 2013.

[101] K. Y. Song and Y. H. Kim, "Characterization of stimulated Brillouin scattering in a few-mode fiber," *Optics Letters*, vol. 38, no. 22, pp. 4841–4844, 2013.

[102] K. Y. Song, Y. H. Kim, and B. Y. Kim, "Intermodal stimulated Brillouin scattering in two-mode fibers," *Optics Letters*, vol. 38, no. 11, pp. 1805–1807, 2013.

[103] V. I. Kovalev and R. G. Harrison, "Waveguide-induced inhomogeneous spectral broadening of stimulated Brillouin scattering in optical fiber," *Optics Letters*, vol. 27, no. 22, pp. 2022–2024, 2002.

[104] H. Eichler, A. Mocofanescu, T. Riesbeck, E. Risse, and D. Bedau, "Stimulated Brillouin scattering in multimode fibers for optical phase conjugation," *Optics Communications*, vol. 208, no. 4-6, pp. 427–431, 2002.

[105] A. Mocofanescu, L. Wang, R. Jain, K. Shaw, A. Gavrielides, P. Peterson, and M. Sharma, "SBS threshold for single mode and multimode GRIN fibers in an all fiber configuration," *Optics Express*, vol. 13, no. 6, pp. 2019–2024, 2005.

[106] G. A. Decker Jr, "Method of fabricating an optical attenuator by fusion splicing of optical fibers," 1985, US Patent 4,557,556.

[107] Y. Koyamada, S. Sato, S. Nakamura, H. Sotobayashi, and W. Chujo, "Simulating and designing Brillouin gain spectrum in single-mode fibers," *Journal of Lightwave Technology*, vol. 22, no. 2, pp. 631–639, 2004.

[108] A. Minardo, R. Bernini, and L. Zeni, "Experimental and numerical study on stimulated Brillouin scattering in a graded-index multimode fiber," *Optics Express*, vol. 22, no. 14, pp. 17 480–17 489, 2014.

[109] M. C. Hudson, "Calculation of the maximum optical coupling efficiency into multimode optical waveguides," *Applied Optics*, vol. 13, no. 5, pp. 1029–1033, 1974.

[110] Y. Koike and A. Inoue, "High-speed graded-index plastic optical fibers and their simple interconnects for 4K/8K video transmission," *Journal of Lightwave Technology*, vol. 34, no. 6, pp. 1551–1555, 2016.

[111] Y. Mizuno and K. Nakamura, "Core alignment of butt coupling between single-mode and multimode optical fibers by monitoring Brillouin scattering signal," *Journal of Lightwave Technology*, vol. 29, no. 17, pp. 2616–2620, 2011.

[112] O. Ziemann, J. Krauser, P. E. Zamzow, and W. Daum, *POF Handbook*, 2nd ed. Springer Berlin Heidelberg, 2008. ISBN 978-3-540-76628-5

[113] M. Olivero, G. Perrone, and A. Vallan, "Near-field measurements and mode power distribution of multimode optical fibers," *IEEE Transactions on Instrumentation and Measurement*, vol. 59, no. 5, pp. 1382–1388, 2010.

[114] D. Marcuse, "Coupled mode theory of round optical fibers," *Bell Labs Technical Journal*, vol. 52, no. 6, pp. 817–842, 1973.

[115] D. Marcuse, "Gaussian approximation of the fundamental modes of graded-index fibers," *JOSA*, vol. 68, no. 1, pp. 103–109, 1978.

[116] A. W. Snyder and R. A. Sammut, "Fundamental (HE 11) modes of graded optical fibers," *Journal of the Optical Society of America A*, vol. 69, no. 12, pp. 1663–1671, 1979.

[117] A. Ghatak and K. Thyagarajan, *An Introduction to Fiber Optics.* Cambridge university press, 1998.

[118] T. P. H. Jeschke, *Untersuchung der Steckerverbindung zwischen polymeroptischen- und quartzglasbasierten Fasern.* Institut für Hochfrequenztechnik Photonik, Technische Universität Berlin, 05/2017.

[119] A. Schreier, T. Jeschke, K. Petermann, A. Wosniok, and K. Krebber, "Analytical model for mode based insertion loss in ball-lensed coupling of graded-index silica and polymer fibres," *26th International Conference on Plastic Optical Fibres, POF 2017 - Proceedings*, 2017.

[120] Corning® SMF-28® ultra optical fiber - product information. [Online]. Available: https://www.corning.com/media/worldwide/coc/documents/Fiber/ SMF-28%20Ultra.pdf

[121] D. Gloge and E. Marcatili, "Multimode theory of grade core fibers," *Bell Labs Technical Journal*, vol. 52, no. 9, pp. 1563–1578, 1973.

[122] D. Marcuse, "Loss analysis of single mode fiber splices," *Bell Labs Technical Journal*, vol. 56, no. 5, pp. 703–718, 1977.

[123] R. Abram, R. Allen, and R. Goodfellow, "The coupling of light-emitting diodes to optical fibers using sphere lenses," *Journal of Applied Physics*, vol. 46, no. 8, pp. 3468–3474, 1975.

[124] J. C. Palais, "Fiber coupling using graded-index rod lenses," *Applied Optics*, vol. 19, no. 12, pp. 2011–2018, 1980.

[125] R. Ulrich and S. Rashleigh, "Beam-to-fiber coupling with low standing wave ratio," *Applied Optics*, vol. 19, no. 14, pp. 2453–2456, 1980.

[126] Y. Koike, H. Suzuki, H. Muto, A. Mitsui, N. Moriya, T. Torikai, and T. Yamauchi, "Optical collimator and optical connector using same," Mar. 3 2015, US Patent 8,967,880.

[127] LaCroix Dynamic Material Selection Data Tool vJanuary 2015. [Online]. Available: https://web.archive.org/web/20151011033820/http: //www.lacroixoptical.com/sites/default/files/content/LaCroix%20Dynamic% 20Material%20Selection%20Data%20Tool%20vJanuary%202015.xlsm

[128] A. Theodosiou, M. Komodromos, and K. Kalli, "Carbon cantilever beam health inspection using a polymer fiber Bragg grating array," *Journal of Lightwave Technology*, vol. 36, no. 4, pp. 986–992, 2018.

[129] Y. Mizuno, N. Hayashi, and K. Nakamura, "Simple coupling method for enhancing Brillouin scattering signal in polymer optical fibres," *Electronics Letters*, vol. 48, no. 20, pp. 1300–1301, 2012.

[130] TIA Standard FOCIS 4, *Fiber Optic Connector Intermateability Standard- Type FC and FC-APC, TIA-604-4-B.* Telecommunications Industry Association, 09-01-2004.

[131] M. Barnoski and S. Jensen, "Fiber waveguides: a novel technique for investigating attenuation characteristics," *Applied Optics*, vol. 15, no. 9, pp. 2112–2115, 1976.

[132] T. Horiguchi and M. Tateda, "BOTDA - Non destructive measurement of single-mode optical fiber attenuation characteristics using Brillouin interaction: Theory," *Journal of Lightwave Technology*, vol. 7, no. 8, pp. 1170–1176, 1989.

[133] V. Lecoeuche, D. J. Webb, C. N. Pannell, and D. A. Jackson, "Transient response in high-resolution Brillouin-based distributed sensing using probe pulses shorter than the acoustic relaxation time," *Optics Letters*, vol. 25, no. 3, pp. 156–158, 2000.

[134] A. Zadok, E. Zilka, A. Eyal, L. Thévenaz, and M. Tur, "Vector analysis of stimulated Brillouin scattering amplification in standard single-mode fibers," *Optics Express*, vol. 16, no. 26, pp. 21 692–21 707, 2008.

[135] O. Shlomovits, T. Langer, and M. Tur, "The effect of source phase noise on stimulated Brillouin amplification," *Journal of Lightwave Technology*, vol. 33, no. 12, pp. 2639–2645, 2015.

[136] A. Minardo, R. Bernini, and L. Zeni, "Analysis of SNR penalty in Brillouin optical time-domain analysis sensors induced by laser source phase noise," *Journal of Optics*, vol. 18, no. 2, p. 025601, 2015.

[137] M. Alem, M. A. Soto, and L. Thévenaz, "Analytical model and experimental verification of the critical power for modulation instability in optical fibers," *Optics Express*, vol. 23, no. 23, pp. 29 514–29 532, 2015.

[138] R. Ruiz-Lombera, J. Urricelqui, M. Sagues, J. Mirapeix, J. M. Lopez-Higuera, and A. Loayssa, "Overcoming nonlocal effects and Brillouin threshold limitations in Brillouin optical time-domain sensors," *IEEE Photonics Journal*, vol. 7, no. 6, pp. 1–9, 2015.

[139] L. Thévenaz, S. F. Mafang, and J. Lin, "Effect of pulse depletion in a Brillouin optical time-domain analysis system," *Optics Express*, vol. 21, no. 12, pp. 14 017–14 035, 2013.

[140] F. Gyger, E. Rochat, S. Chin, M. Niklès, and L. Thévenaz, "Extending the sensing range of Brillouin optical time-domain analysis up to 325 km combining four optical repeaters," in *23rd International Conference on Optical Fibre Sensors*, vol. 9157. International Society for Optics and Photonics, 2014, p. 91576Q.

[141] M. Taki, Y. Muanenda, C. Oton, T. Nannipieri, A. Signorini, and F. Di Pasquale, "Cyclic pulse coding for fast BOTDA fiber sensors," *Optics Letters*, vol. 38, no. 15, pp. 2877–2880, 2013.

[142] A. Dominguez-Lopez, A. Lopez-Gil, S. Martin-Lopez, and M. Gonzalez-Herraez, "Strong cancellation of RIN transfer in a Raman-assisted BOTDA using balanced detection," *IEEE Photonics Technology Letters*, vol. 26, no. 18, pp. 1817–1820, 2014.

[143] A. Zornoza, M. Sagues, and A. Loayssa, "Self-heterodyne detection for SNR improvement and distributed phase-shift measurements in BOTDA," *journal of Lightwave Technology*, vol. 30, no. 8, pp. 1066–1072, 2011.

[144] A. W. Brown, B. G. Colpitts, and K. Brown, "Dark-pulse Brillouin optical time-domain sensor with 20-mm spatial resolution," *Journal of Lightwave Technology*, vol. 25, no. 1, pp. 381–386, 2007.

[145] T. Sperber, A. Eyal, M. Tur, and L. Thévenaz, "High spatial resolution distributed sensing in optical fibers by Brillouin gain-profile tracing," *Optics Express*, vol. 18, no. 8, pp. 8671–8679, 2010.

[146] W. Li, X. Bao, Y. Li, and L. Chen, "Differential pulse-width pair BOTDA for high spatial resolution sensing," *Optics Express*, vol. 16, no. 26, pp. 21 616–21 625, 2008.

[147] K. Kishida and C. Li, "Pulse pre-pump-BOTDA technology for new generation of distributed strain measuring system," *Structural health monitoring and intelligent infrastructure*, vol. 1, pp. 471–477, 2005.

[148] X. Bao, A. Brown, M. DeMerchant, and J. Smith, "Characterization of the Brillouin-loss spectrum of single-mode fibers by use of very short pulses," *Optics Letters*, vol. 24, no. 8, pp. 510–512, 1999.

[149] M. S. D. B. Zan, T. Tsumuraya, and T. Horiguchi, "The use of Walsh code in modulating the pump light of high spatial resolution phase-shift-pulse Brillouin optical time domain analysis with non-return-to-zero pulses," *Measurement Science and Technology*, vol. 24, no. 9, p. 094025, 2013.

[150] R. Bernini, A. Minardo, and L. Zeni, "Dynamic strain measurement in optical fibers by stimulated Brillouin scattering," *Optics Letters*, vol. 34, no. 17, pp. 2613–2615, 2009.

[151] A. Voskoboinik, O. F. Yilmaz, A. W. Willner, and M. Tur, "Sweep-free distributed Brillouin time-domain analyzer (SF-BOTDA)," *Optics Express*, vol. 19, no. 26, pp. B842–B847, Dec 2011.

[152] Q. Bai, Q. Wang, D. Wang, Y. Wang, Y. Gao, H. Zhang, M. Zhang, and B. Jin, "Recent advances in Brillouin optical time domain reflectometry," *Sensors*, vol. 19, no. 8, p. 1862, 2019.

[153] Y. Koyamada, Y. Sakairi, N. Takeuchi, and S. Adachi, "Novel technique to improve spatial resolution in Brillouin optical time-domain reflectometry," *IEEE Photonics Technology Letters*, vol. 19, no. 23, pp. 1910–1912, 2007.

[154] K. Nishiguchi, C.-H. Li, A. Guzik, and K. Kishida, "Synthetic spectrum approach for Brillouin optical time-domain reflectometry," *Sensors*, vol. 14, no. 3, pp. 4731–4754, 2014.

[155] R. Shibata, H. Kasahara, L. P. Elias, and T. Horiguchi, "Improving performance of phase shift pulse BOTDR," *IEICE Electronics Express*, pp. 14–20 170 267, 2017.

[156] M. S. D. Zan, Y. Masui, and T. Horiguchi, "Differential cross spectrum technique for improving the spatial resolution of BOTDR sensor," in *2018 IEEE 7th International Conference on Photonics (ICP)*. IEEE, 2018, pp. 1–3.

[157] Y. Peled, A. Motil, L. Yaron, and M. Tur, "Slope-assisted fast distributed sensing in optical fibers with arbitrary Brillouin profile," *Optics Express*, vol. 19, no. 21, pp. 19 845–19 854, 2011.

[158] D. Maraval, R. Gabet, Y. Jaouën, and V. Lamour, "Slope-assisted BOTDR for pipeline vibration measurements," in *2017 25th Optical Fiber Sensors Conference (OFS)*. IEEE, 2017, pp. 1–4.

[159] A. Motil, O. Danon, Y. Peled, and M. Tur, "Pump-power-independent double slope-assisted distributed and fast Brillouin fiber-optic sensor," *IEEE Photonics Technology Letters*, vol. 26, no. 8, pp. 797–800, 2014.

[160] D. Zhou, Y. Dong, B. Wang, T. Jiang, D. Ba, P. Xu, H. Zhang, Z. Lu, and H. Li, "Slope-assisted BOTDA based on vector SBS and frequency-agile technique for wide-strain-range dynamic measurements," *Optics Express*, vol. 25, no. 3, pp. 1889–1902, 2017.

[161] K.-Y. Song, Z. He, and K. Hotate, "Distributed strain measurement with millimeter-order spatial resolution based on brillouin optical correlation domain analysis and beat lock-in detection scheme," in *Optical Fiber Sensors*. Optical Society of America, 2006, p. ThC2.

[162] K. Hotate and T. Hasegawa, "Measurement of Brillouin gain spectrum distribution along an optical fiber using a correlation-based technique–proposal, experiment and simulation," *IEICE Transactions on Electronics*, vol. 83, no. 3, pp. 405–412, 2000.

[163] K. Hotate, "Fiber distributed Brillouin sensing with optical correlation domain techniques," *Optical Fiber Technology*, vol. 19, no. 6, pp. 700–719, 2013.

[164] L. Thévenaz, *Advanced Fiber Optics*. EPFL Press, 2011.

[165] K. Hotate, H. Arai, and K. Y. Song, "Range-enlargement of simplified Brillouin optical correlation domain analysis based on a temporal gating scheme," *SICE Journal of Control, Measurement, and System Integration*, vol. 1, no. 4, pp. 271–274, 2008.

[166] K. Y. Song, M. Kishi, Z. He, and K. Hotate, "High-repetition-rate distributed Brillouin sensor based on optical correlation-domain analysis with differential frequency modulation," *Optics Letters*, vol. 36, no. 11, pp. 2062–2064, 2011.

[167] C. Zhang, M. Kishi, and K. Hotate, "5,000 points/s high-speed random accessibility for dynamic strain measurement at arbitrary multiple points along a fiber by Brillouin optical correlation domain analysis," *Applied Physics Express*, vol. 8, no. 4, p. 042501, 2015.

[168] J. H. Jeong, K. Lee, K. Y. Song, J.-M. Jeong, and S. B. Lee, "Differential measurement scheme for Brillouin optical correlation domain analysis," *Optics Express*, vol. 20, no. 24, pp. 27 094–27 101, 2012.

[169] Y. Antman, N. Primerov, J. Sancho, L. Thévenaz, and A. Zadok, "Localized and stationary dynamic gratings via stimulated Brillouin scattering with phase modulated pumps," *Optics Express*, vol. 20, no. 7, pp. 7807–7821, 2012.

[170] Y. Antman, L. Yaron, T. Langer, M. Tur, N. Levanon, and A. Zadok, "Experimental demonstration of localized Brillouin gratings with low off-peak reflectivity established by perfect Golomb codes," *Optics Letters*, vol. 38, no. 22, pp. 4701–4704, 2013.

[171] S. Manotham, M. Kishi, Z. He, and K. Hotate, "1-cm spatial resolution with large dynamic range in strain distributed sensing by Brillouin optical correlation domain reflectometry based on intensity modulation," in *Third Asia Pacific Optical Sensors Conference*, vol. 8351. International Society for Optics and Photonics, 2012, p. 835136.

[172] B. Wang, X. Fan, Y. Fu, and Z. He, "Distributed dynamic strain measurement based on dual-slope-assisted Brillouin optical correlation domain analysis," *Journal of Lightwave Technology*, 2019.

[173] K.-Y. Song and K. Hotate, "Brillouin optical correlation domain analysis in linear configuration," *IEEE Photonics Technology Letters*, vol. 20, no. 24, pp. 2150–2152, 2008.

[174] Y. Mizuno, Z. He, and K. Hotate, "Measurement range enlargement in Brillouin optical correlation-domain reflectometry based on temporal gating scheme," *Optics Express*, vol. 17, no. 11, pp. 9040–9046, 2009.

[175] H. Lee, N. Hayashi, Y. Mizuno, and K. Nakamura, "Slope-assisted Brillouin optical correlation-domain reflectometry: proof of concept," *IEEE Photonics Journal*, vol. 8, no. 3, pp. 1–7, 2016.

[176] H. Lee, Y. Mizuno, and K. Nakamura, "Measurement sensitivity dependencies on incident power and spatial resolution in slope-assisted Brillouin optical correlation-domain reflectometry," *Sensors and Actuators A: Physical*, vol. 268, pp. 68–71, 2017.

[177] H. Lee, K. Noda, Y. Mizuno, and K. Nakamura, "Trade-off relation between strain dynamic range and spatial resolution in slope-assisted Brillouin optical correlation-domain reflectometry," *Measurement Science and Technology*, vol. 30, no. 7, p. 075204, 2019.

[178] H. Lee, Y. Mizuno, and K. Nakamura, "Enhanced stability and sensitivity of slope-assisted Brillouin optical correlation-domain reflectometry using polarization-maintaining fibers," *OSA Continuum*, vol. 2, no. 3, pp. 874–880, 2019.

[179] D. Garus, T. Gogolla, K. Krebber, and F. Schliep, "Distributed sensing technique based on Brillouin optical-fiber frequency-domain analysis," *Optics Letters*, vol. 21, no. 17, pp. 1402–1404, 1996. doi: dx.doi.org/10.1364/OL.21.001402

[180] A. Wosniok, "Distributed Brillouin Sensing: Frequency-Domain Techniques," *Handbook of Optical Fibers*, pp. 1–25, 2017.

[181] D. Gottlieb and C.-W. Shu, "On the Gibbs phenomenon and its resolution," *SIAM Review*, vol. 39, no. 4, pp. 644–668, 1997.

[182] R. Bernini, A. Minardo, and L. Zeni, "Distributed sensing at centimeter-scale spatial resolution by BOFDA: Measurements and signal processing," *IEEE Photonics Journal*, vol. 4, no. 1, pp. 48–56, 2011.

[183] T. Gogolla, *Theoretische Untersuchung der Brillouin-Wechselwirkung in Lichtleitfasern zur kontinuierlich verteilten Temperatur-und Dehnungsmessung auf Basis der Frequenzbereichsanalyse.* Shaker, 2000.

[184] L. Zeni, E. Catalano, A. Coscetta, and A. Minardo, "High-pass filtering for accurate reconstruction of the Brillouin frequency shift profile from Brillouin optical frequency domain analysis data," *IEEE Sensors Journal*, vol. 18, no. 1, pp. 185–192, 2017.

[185] T. Kapa, A. Schreier, and K. Krebber, "63 km BOFDA for temperature and strain monitoring," *Sensors*, vol. 18, no. 5, p. 1600, 2018.

[186] K. George, C.-I. H. Chen, and J. B. Tsui, "Extension of two-signal spurious-free dynamic range of wideband digital receivers using Kaiser window and compensation method," *IEEE Transactions on Microwave Theory and Techniques*, vol. 55, no. 4, pp. 788–794, 2007.

[187] A. Minardo, A. Coscetta, E. Catalano, R. Bernini, and L. Zeni, "High spatial resolution physical and chemical sensing based on BOFDA," in *Optical Sensors 2019*, vol. 11028. International Society for Optics and Photonics, 2019, p. 110281B.

[188] A. Wosniok, Y. Mizuno, K. Krebber, and K. Nakamura, "L-BOFDA: a new sensor technique for distributed Brillouin sensing," in *Fifth European Workshop on Optical Fibre Sensors*, vol. 8794. International Society for Optics and Photonics, 2013, p. 879431.

[189] R. Ruiz-Lombera, A. Minardo, R. Bernini, J. Mirapeix, J. M. Lopez-Higuera, and L. Zeni, "Experimental demonstration of a Brillouin optical frequency-domain reflectometry (BOFDR) sensor," in *2017 25th Optical Fiber Sensors Conference (OFS)*. IEEE, 2017, pp. 1–4.

[190] A. Y. Mohamed and N. Nöther, "Novel realization of Brillouin optical frequency domain reflectometry (BOFDR) with digital envelope detection," in *Optical Fiber Sensors*. Optical Society of America, 2018, p. TuE22.

[191] N. Hayashi, Y. Mizuno, D. Koyama, and K. Nakamura, "Measurement of acoustic velocity in poly (methyl methacrylate)-based polymer optical fiber for Brillouin frequency shift estimation," *Applied Physics Express*, vol. 4, no. 10, p. 102501, 2011.

[192] N. Hayashi, Y. Mizuno, D. Koyama, and K. Nakamura, "Dependence of Brillouin frequency shift on temperature and strain in poly (methyl methacrylate)-based polymer optical fibers estimated by acoustic velocity measurement," *Applied Physics Express*, vol. 5, no. 3, p. 032502, 2012.

[193] K. Minakawa, K. Koike, N. Hayashi, Y. Koike, Y. Mizuno, and K. Nakamura, "Dependence of Brillouin frequency shift on water absorption ratio in polymer optical fibers," *Journal of Applied Physics*, vol. 119, no. 22, p. 223102, 2016.

[194] K. Minakawa, N. Hayashi, Y. Mizuno, and K. Nakamura, "Potential applicability of Brillouin scattering in partially chlorinated polymer optical fibers to high-precision temperature sensing," *Applied Physics Express*, vol. 6, no. 5, p. 052501, 2013.

[195] Y. Mizuno, T. Ishigure, and K. Nakamura, "Brillouin gain spectrum characterization in perfluorinated graded index polymer optical fiber with 62.5 um core diameter," *IEEE Photonics Technology Letters*, vol. 23, no. 24, pp. 1863–1865, 2011.

[196] K. Minakawa, N. Hayashi, Y. Shinohara, M. Tahara, H. Hosoda, Y. Mizuno, and K. Nakamura, "Wide-range temperature dependences of Brillouin scattering properties in polymer optical fiber," *Japanese Journal of Applied Physics*, vol. 53, no. 4, p. 042502, 2014.

[197] N. Hayashi, K. Minakawa, Y. Mizuno, and K. Nakamura, "Brillouin frequency shift hopping in polymer optical fiber," *Applied Physics Letters*, vol. 105, no. 9, p. 091113, 2014.

[198] Y. Mizuno, N. Matsutani, N. Hayashi, H. Lee, M. Tahara, H. Hosoda, and K. Nakamura, "Brillouin characterization of slimmed polymer optical fibers for strain sensing with extremely wide dynamic range," *Optics Express*, vol. 26, no. 21, pp. 28030–28037, 2018.

[199] K. Minakawa, Y. Mizuno, and K. Nakamura, "Cross effect of strain and temperature on Brillouin frequency shift in polymer optical fibers," *Journal of Lightwave Technology*, vol. 35, no. 12, pp. 2481–2486, Jun 2017.

[200] Y. Dong, P. Xu, H. Zhang, Z. Lu, L. Chen, and X. Bao, "Characterization of evolution of mode coupling in a graded index polymer optical fiber by using Brillouin optical time domain analysis," *Optics Express*, vol. 22, no. 22, pp. 26510–26516, 2014.

[201] Y. Mizuno, N. Hayashi, and K. Nakamura, "Brillouin scattering signal in polymer optical fiber enhanced by exploiting pulsed pump with multimode-fiber-assisted coupling technique," *Optics Letters*, vol. 38, no. 9, pp. 1467–1469, 2013.

[202] Y. Mizuno, N. Hayashi, H. Tanaka, K. Nakamura, and S.-i. Todoroki, "Propagation mechanism of polymer optical fiber fuse," *Scientific Reports*, vol. 4, p. 4800, 2014.

[203] T. Kurashima, T. Horiguchi, and M. Tateda, "Thermal effects on the Brillouin frequency shift in jacketed optical silica fibers," *Applied optics*, vol. 29, no. 15, pp. 2219–2222, 1990.

[204] T. Horiguchi, T. Kurashima, and M. Tateda, "Tensile strain dependence of Brillouin frequency shift in silica optical fibers," *IEEE Photonics Technology Letters*, vol. 1, no. 5, pp. 107–108, 1989.

[205] W. Zou, Z. He, M. Kishi, and K. Hotate, "Stimulated Brillouin scattering and its dependences on temperature and strain in a high-delta optical fiber with F-doped depressed inner-cladding," in *Optical Fiber Sensors*. Optical Society of America, 2006, p. ThE38.

[206] W. Zou, Z. He, and K. Hotate, "Investigation of strain and temperature dependences of Brillouin frequency shifts in GeO2-doped optical fibers," *Journal of Lightwave Technology*, vol. 26, no. 13, pp. 1854–1861, 2008.

[207] A. Wosniok and A. Schreier, "Toward investigation of brillouin scattering in multi-mode polymer and silica optical fibers," in *SPIE Optics+ Optoelectronics*. International Society for Optics and Photonics, 2017, pp. 102320M–102320M.

[208] K. Krebber, "Ortsauflösende Lichtleitfaser-Sensorik für Temperatur und Dehnung unter Nutzung der stimulierten Brillouin-Streuung basierend auf der Frequenzbereichsanalyse," Ph.D. dissertation, Ruhr-Universität Bochum, Fakultät für Elektrotechnik und Informationstechnik, 2001.

[209] K. E. Carroll, C. Zhang, D. J. Webb, K. Kalli, A. Argyros, and M. C. Large, "Thermal response of Bragg gratings in PMMA microstructured optical fibers," *Optics Express*, vol. 15, no. 14, pp. 8844–8850, 2007.

[210] M. A. Soto and L. Thévenaz, "Modeling and evaluating the performance of Brillouin distributed optical fiber sensors," *Optics Express*, vol. 21, no. 25, pp. 31347–31366, 2013.

[211] T. Strutz, *Data Fitting and Uncertainty: A Practical Introduction to Weighted Least Squares and Beyond.* Vieweg and Teubner, 2010.

[212] K. Levenberg, "A method for the solution of certain non-linear problems in least squares," *Quarterly of Applied Mathematics*, vol. 2, no. 2, pp. 164–168, 1944.

[213] D. W. Marquardt, "An algorithm for least-squares estimation of nonlinear parameters," *Journal of the Society for Industrial and Applied Mathematics*, vol. 11, no. 2, pp. 431–441, 1963.

[214] A. W. Brown, B. G. Colpitts, and K. Brown, "Distributed sensor based on dark-pulse Brillouin scattering," *Photonics Technology Letters, IEEE*, vol. 17, no. 7, pp. 1501–1503, 2005.

[215] H. Liang, W. Li, N. Linze, L. Chen, and X. Bao, "High-resolution DPP-BOTDA over 50 km LEAF using return-to-zero coded pulses," *Opt. Lett.*, vol. 35, no. 10, pp. 1503–1505, May 2010.

[216] Passive optical components – optical circulator. [Online]. Available: www.fiberopticshare.com/passive-optical-components-optical-circulator.html

[217] D. Marcuse, *Theory of Dielectric Optical Waveguides.* Elsevier, 2013.

[218] M. C. Simon, "Ray tracing formulas for monoaxial optical components," *Applied optics*, vol. 22, no. 2, pp. 354–360, 1983.

[219] M. Avendaño-Alejo, "Analysis of the refraction of the extraordinary ray in a plane-parallel uniaxial plate with an arbitrary orientation of the optical axis." *Optics express*, vol. 13, no. 7, pp. 2549–2555, 2005.

[220] O. A. Alduchov and R. E. Eskridge, "Improved magnus form approximation of saturation vapor pressure," *Journal of Applied Meteorology*, vol. 35, no. 4, pp. 601–609, 1996.

[221] J. G. Speight *et al.*, *Lange's Handbook of Chemistry.* McGraw Hill New York, 2005, vol. 1.

[222] M. R. Moldover, J. M. Trusler, T. Edwards, J. B. Mehl, and R. S. Davis, "Measurement of the universal gas constant R using a spherical acoustic resonator," *Physical Review Letters*, vol. 60, no. 4, p. 249, 1988.

[223] M. Nikles, L. Thevenaz, and P. A. Robert, "Brillouin gain spectrum characterization in single mode optical fibers," *Journal of Lightwave Technology*, vol. 15, no. 10, pp. 1842–1851, 1997.

[224] A. Shenoy, *Thermoplastic Melt Rheology and Processing.* CRC Press, 1996.

[225] L. Rothman, I. Gordon, A. Barbe, D. Benner, P. Bernath, M. Birk, V. Boudon, L. Brown, A. Campargue, J.-P. Champion, K. Chance, L. Coudert, V. Dana, V. Devi, S. Fally, J.-M. Flaud, R. Gamache, A. Goldman, D. Jacquemart, I. Kleiner, N. Lacome, W. Lafferty, J.-Y. Mandin, S. Massie, S. Mikhailenko, C. Miller, N. Moazzen-Ahmadi, O. Naumenko, A. Nikitin, J. Orphal, V. Perevalov, A. Perrin, A. Predoi-Cross, C. Rinsland, M. Rotger, M. Šimečková, M. Smith, K. Sung, S. Tashkun, J. Tennyson, R. Toth, A. Vandaele, and J. V. Auwera, "The HITRAN 2008 molecular spectroscopic database," *Journal of Quantitative Spectroscopy and Radiative Transfer*, vol. 110, no. 9, pp. 533 – 572, 2009, hITRAN.

[226] D. O'Hagan, "Understanding organofluorine chemistry. An introduction to the C–F bond," *Chemical Society Reviews*, vol. 37, pp. 308–319, 2008.

[227] S. Spießberger, M. Schiemangk, A. Wicht, H. Wenzel, G. Erbert, and G. Tränkle, "DBR laser diodes emitting near 1064 nm with a narrow intrinsic linewidth of 2 kHz," *Applied Physics B*, vol. 104, no. 4, p. 813, Jul 2011.

[228] A. Klehr, B. Sumpf, K.-H. Hasler, J. Fricke, A. Liero, and G. Erbert, "High-power pulse generation in GHz range with 1064 nm DBR tapered laser," *IEEE Photonics Technology Letters*, vol. 22, no. 11, pp. 832–834, 2010.

[229] P. Atkins and J. de Paula, *Atkins' Physical Chemistry*. Oxford University Press, 2006.

[230] O. Reshef, K. Shtyrkova, M. G. Moebius, S. Griesse-Nascimento, S. Spector, C. C. Evans, E. Ippen, and E. Mazur, "Polycrystalline anatase titanium dioxide microring resonators with negative thermo-optic coefficient," *JOSA B*, vol. 32, no. 11, pp. 2288–2293, 2015.

[231] Y. J. Weitsman, *Fluid Effects in Polymers and Polymeric Composites*. Springer Science & Business Media, 2011.

[232] H. Bair, G. Johnson, and R. Merriweather, "Water sorption of polycarbonate and its effect on the polymer's dielectric behavior," *Journal of Applied Physics*, vol. 49, no. 10, pp. 4976–4984, 1978.

[233] Y. Luo, B. Yan, Q. Zhang, G.-D. Peng, J. Wen, and J. Zhang, "Fabrication of polymer optical fibre (POF) gratings," *Sensors*, vol. 17, no. 3, p. 511, 2017.

[234] W. Yuan, A. Stefani, M. Bache, T. Jacobsen, B. Rose, N. Herholdt-Rasmussen, F. K. Nielsen, S. Andresen, O. B. Sørensen, K. S. Hansen, and O. Bang, "Improved thermal and strain performance of annealed polymer optical fiber Bragg gratings," *Optics Communications*, vol. 284, no. 1, pp. 176–182, 2011.

[235] J. Urricelqui, M. A. Soto, and L. Thévenaz, "Sources of noise in brillouin optical time-domain analyzers," in *International Conference on Optical Fibre Sensors (OFS24)*. International Society for Optics and Photonics, 2015. doi: dx.doi.org/10.1117/12.2195298 pp. 963 434–963 434.

[236] T. Horiguchi, K. Shimizu, T. Kurashima, M. Tateda, and Y. Koyamada, "Development of a distributed sensing technique using Brillouin scattering," *Journal of Lightwave Technology*, vol. 13, no. 7, pp. 1296–1302, 1995.

[237] P. Kronenberg, P. K. Rastogi, P. Giaccari, and H. G. Limberger, "Relative humidity sensor with optical fiber Bragg gratings," *Optics Letters*, vol. 27, no. 16, pp. 1385–1387, 2002.

[238] C. Galindez, F. J. Madruga, and J. M. Lopez-Higuera, "Influence of humidity on the measurement of Brillouin frequency shift," *IEEE Photonics Technology Letters*, vol. 20, no. 23, pp. 1959–1961, 2008.

[239] International Electrotechnical Commission, *Temperature measurement - Distributed sensing (IEC 61757-2: 2016)*. International Electrotechnical Commission, 2016.

[240] Y. Yu, L. Luo, B. Li, K. Soga, and J. Yan, "Frequency resolution quantification of Brillouin-distributed optical fiber sensors," *IEEE Photonics Technology Letters*, vol. 28, no. 21, pp. 2367–2370, 2016.

[241] H. Zheng, Z. Fang, Z. Wang, B. Lu, Y. Cao, Q. Ye, R. Qu, and H. Cai, "Brillouin frequency shift of fiber distributed sensors extracted from noisy signals by quadratic fitting," *Sensors*, vol. 18, no. 2, p. 409, 2018.

[242] A. Azad, L. Wang, N. Guo, C. Lu, and H. Tam, "Temperature sensing in BOTDA system by using artificial neural network," *Electronics Letters*, vol. 51, no. 20, pp. 1578–1580, 2015.

[243] A. K. Azad, L. Wang, N. Guo, H.-Y. Tam, and C. Lu, "Signal processing using artificial neural network for BOTDA sensor system," *Optics Express*, vol. 24, no. 6, pp. 6769–6782, 2016.

[244] R. Ruiz-Lombera, A. Fuentes, L. Rodriguez-Cobo, J. M. Lopez-Higuera, and J. Mirapeix, "Simultaneous temperature and strain discrimination in a conventional BOTDA via artificial neural networks," *Journal of Lightwave Technology*, vol. 36, no. 11, pp. 2114–2121, 2018.

[245] H. Lee, Y. Mizuno, and K. Nakamura, "Enhanced stability and sensitivity of slope-assisted Brillouin optical correlation-domain reflectometry using polarization-maintaining fibers," *OSA Continuum*, vol. 2, no. 3, pp. 874–880, 2019.

[246] R. Ruiz-Lombera, I. Laarossi, L. Rodríguez-Cobo, M. Á. Quintela, J. M. López-Higuera, and J. Mirapeix, "Distributed high-temperature optical fiber sensor based on a Brillouin optical time domain analyzer and multimode gold-coated fiber," *IEEE Sensors Journal*, vol. 17, no. 8, pp. 2393–2397, 2017.

[247] Y. Dong, L. Chen, and X. Bao, "Time-division multiplexing-based BOTDA over 100km sensing length," *Optics Letters*, vol. 36, no. 2, pp. 277–279, 2011.

[248] H. Ohno, H. Naruse, N. Yasue, Y. Miyajima, H. Uchiyama, Y. Sakairi, and Z. X. Li, "Development of highly stable BOTDR strain sensor employing microwave heterodyne detection and tunable electric oscillator," in *Advanced Photonic Sensors and Applications II*, vol. 4596. International Society for Optics and Photonics, 2001, pp. 74–86.

[249] K. Hotate and H. Arai, "Enlargement of measurement range of simplified BOCDA fiber-optic distributed strain sensing system using a temporal gating scheme," in *17th International Conference on Optical Fibre Sensors*, vol. 5855. International Society for Optics and Photonics, 2005, pp. 184–188.

[250] J. Fang, M. Sun, D. Che, M. Myers, H. Bao, C. Prohasky, and W. Shieh, "Complex Brillouin optical time-domain analysis," *Journal of Lightwave Technology*, vol. 36, no. 10, pp. 1840–1850, 2018.

[251] P. Walters, *An Introduction to Ergodic Theory.* Springer Science & Business Media, 2000, vol. 79.

[252] P. Richter, "Estimating errors in least-squares fitting," *The Telecommunications and Data Acquisition Progress Report 42-122*, p. 107, 1995.

[253] M. A. Soto, J. A. Ramirez, and L. Thévenaz, "Intensifying the response of distributed optical fibre sensors using 2D and 3D image restoration," *Nature Communications*, vol. 7, p. 10870, 2016.

[254] H. S. Pradhan and P. Sahu, "Brillouin distributed strain sensor performance improvement using FourWaRD algorithm," *Optik-International Journal for Light and Electron Optics*, vol. 127, no. 5, pp. 2666–2669, 2016.

[255] W. Zou, Z. He, and K. Hotate, "Experimental study of Brillouin scattering in fluorine-doped single-mode optical fibers," *Opt. Express*, vol. 16, no. 23, pp. 18 804–18 812, Nov 2008.

[256] X. Bao, J. Dhliwayo, N. Heron, D. J. Webb, and D. A. Jackson, "Experimental and theoretical studies on a distributed temperature sensor based on Brillouin scattering," *Journal of Lightwave Technology*, vol. 13, no. 7, pp. 1340–1348, 1995.

[257] R. Bernini, L. Crocco, A. Minardo, F. Soldovieri, and L. Zeni, "All frequency domain distributed fiber-optic Brillouin sensing," *IEEE Sensors Journal*, vol. 3, no. 1, pp. 36–43, 2003.

[258] K. O. Hill and G. Meltz, "Fiber Bragg grating technology fundamentals and overview," *Journal of Lightwave technology*, vol. 15, no. 8, pp. 1263–1276, 1997.

[259] C. Roychoudhuri, "Response of Fabry–Perot interferometers to light pulses of very short duration," *Journal of the Optical Society of America*, vol. 65, no. 12, pp. 1418–1426, 1975.

[260] S. Liehr, N. Nöther, M. Steffen, O. Gili, and K. Krebber, "Performance of digital incoherent OFDR and prospects for optical fiber sensing applications," in *23rd International Conference on Optical Fibre Sensors*, vol. 9157. International Society for Optics and Photonics, 2014, p. 915737.

[261] X. Lu, M. A. Soto, M. G. Herraez, and L. Thévenaz, "Brillouin distributed fibre sensing using phase modulated probe," in *Fifth European Workshop on Optical Fibre Sensors*, vol. 8794. International Society for Optics and Photonics, 2013, p. 87943P.

[262] D. M. Nguyen, B. Stiller, M. W. Lee, J.-C. Beugnot, H. Maillotte, A. Mottet, J. Hauden, and T. Sylvestre, "Distributed Brillouin fiber sensor with enhanced sensitivity based on anti-Stokes single-sideband suppressed-carrier modulation," *IEEE Photonics Technology Letters*, vol. 25, no. 1, pp. 94 – 96, 2013.

[263] A. Schreier, S. Liehr, A. Wosniok, and K. Krebber, "Comparison of solution approaches for distributed humidity sensing in perfluorinated graded-index polymer optical fibers," in *Optical Sensors 2019*, vol. 11028. International Society for Optics and Photonics, 2019, p. 1102808.

[264] S. Liehr, N. Nöther, and K. Krebber, "Incoherent optical frequency domain reflectometry and distributed strain detection in polymer optical fibers," *Measurement Science and Technology*, vol. 21, no. 1, p. 017001, 2009.

[265] A. Theodosiou, K. Kalli, and M. Komodromos, "Health monitoring of carbon cantilever using femtosecond laser inscribed FBG array in gradient-index CYTOP polymer fibre," in *Optical Fiber Sensors Conference (OFS), 2017 25th*. IEEE, 2017, pp. 1–4.

[266] S. Liehr, P. Lenke, M. Wendt, and K. Krebber, "Perfluorinated graded-index polymer optical fibers for distributed measurement of strain," in *17th International Conference on Plastic Optical Fibers (POF 2008)*, 2008.

[267] W. Zhang, D. J. Webb, and G.-D. Peng, "Investigation into time response of polymer fiber Bragg grating based humidity sensors," *Journal of Lightwave Technology*, vol. 30, no. 8, pp. 1090–1096, 2012.

[268] X. Chen, W. Zhang, C. Liu, Y. Hong, and D. J. Webb, "Enhancing the humidity response time of polymer optical fiber Bragg grating by using laser micromachining," *Optics Express*, vol. 23, no. 20, pp. 25 942–25 949, 2015.

List of publications

Due to priority reasons, parts of the results of this thesis have been published in peer-reviewed scientific journals and on internationally renowned conferences. The latest status of this list is available at https://orcid.org/0000-0002-8424-3899.

Journals

1. Kapa, T., Schreier, A., Krebber, K. 63 km BOFDA for Temperature and Strain Monitoring. *Sensors* **2018**, *18, 5*, 1600, https://doi.org/10.3390/s18051600.

2. Schreier, A., Wosniok, A., Liehr, S., Krebber, K. Humidity-induced Brillouin frequency shift in perfluorinated polymer optical fibers. *Optics Express* **2018**, *26, 17*, 22307–22314, https://doi.org/10.1364/OE.26.022307.

3. Schreier, A., Liehr, S., Wosniok, A., Krebber, K. Investigation on the Influence of Humidity on Stimulated Brillouin Backscattering in Perfluorinated Polymer Optical Fibers. *Sensors* **2018**, *18, 11*, 3952, https://doi.org/10.3390/s18113952.

4. Kapa, T., Schreier, A., Krebber, K. A 100-km BOFDA Assisted by First-Order Bi-Directional Raman Amplification. *Sensors* **2019**, *19, 7*, 1527, https://doi.org/10.3390/s19071527.

5. Schreier, A., Kapa, T., Wosniok, A., Kowarik, S. Stochastic modeling of the frequency uncertainty in Brillouin-based fiber sensors. *under review* , **2020**

Conferences

6. Wosniok, A., Schreier, A. Toward investigation of Brillouin scattering in multimode polymer and silica optical fibers. *Proceedings of SPIE* Volume 10232, Microstructured and Specialty Optical Fibres V, 102320M, **2017**, https://doi.org/10.1117/12.2267823.

7. Schreier, A., Jeschke, T., Petermann, K., Wosniok, A., Krebber, K. Analytical model for mode based insertion loss in ball-lensed coupling of graded-index silica and polymer fibres. 26th International Conference on Plastic Optical Fibres, POF 2017 - Proceedings, **2017**.

8. Schreier, A., Liehr, S., Wosniok, A., Krebber, K. Comparison of solution approaches for distributed humidity sensing in perfluorinated graded-index polymer optical fibers. *Proceedings of SPIE* Volume 11028, Optical Sensors, 1102808, **2019**, `https://doi.org/10.1117/12.2522307`.

9. Wosniok, A., Schreier, A., Krebber, K. Towards Distributed structural health monitoring system based on Brillouin Scattering in perfluorinated polymer optical fibers. *28th International Conference on Plastic Optical Fibres*, **2019**